OECD Review of Agricultural Policies: Switzerland 2015

OECD
BETTER POLICIES FOR BETTER LIVES

This work is published on the responsibility of the Secretary-General of the OECD. The opinions expressed and arguments employed herein do not necessarily reflect the official views of the Organisation or of the governments of its member countries.

This document and any map included herein are without prejudice to the status of or sovereignty over any territory, to the delimitation of international frontiers and boundaries and to the name of any territory, city or area.

Please cite this publication as:
OECD (2015), *OECD Review of Agricultural Policies: Switzerland 2015*, OECD Publishing.
http://dx.doi.org/10.1787/9789264168039-en

ISBN 978-92-64-22654-8 (print)
ISBN 978-92-64-16803-9 (PDF)

Series: OECD Review of Agricultural Policies
ISSN 1990-0058 (print)
ISSN 1990-004X (online)

The statistical data for Israel are supplied by and under the responsibility of the relevant Israeli authorities. The use of such data by the OECD is without prejudice to the status of the Golan Heights, East Jerusalem and Israeli settlements in the West Bank under the terms of international law.

Photo credits: © Grigory Fedyukovich/iStock/Thinkstock; © svedoliver/iStock/Thinkstock; © Shaiith/Shutterstock.

Corrigenda to OECD publications may be found on line at: *www.oecd.org/publishing/corrigenda*.

Foreword

The country's four main agricultural policy objectives, set in the federal constitution are: food security (make an essential contribution towards ensuring food supplies for the population); sustainability of agricultural production (production measures which maintain fertile soils and clean drinking water are to be used); taking care of the landscape is seen as an essential task of agriculture; and finally agriculture is to help maintain rural areas. To meet these objectives Switzerland has adopted an elaborate system of agricultural policy measures, combining border measures and direct payments to farmers that add up to a relatively high level of support to its farming sector. Since the mid-1990s Switzerland has been implementing gradual reforms to its farming policies by reducing market interventions and increasing the role of direct payments.

The OECD Secretariat carried the last evaluation of Swiss agricultural policies in the late 1980s (OECD, 1988). During the 1990s and early 2000 Switzerland has carried out a set of reforms of its agricultural policy. Domestic market interventions were gradually discontinued and all state guarantees for prices and sales were abolished and border measures reduced. Within the reforms the amount of direct payments has increased and the system of direct payments fine-tuned. Altogether these policy reforms constitute a gradual but significant change in policies implemented since the mid-1990s.

This study aims to map these policy reforms and their impact on the level and structure of support to agriculture; and to evaluate the policy reforms and make recommendation concerning the continuing reform process. The study uses the PSE/CSE/GSSE data as well as an augmented version of the Policy Evaluation Model (PEM) to assess economic and environmental performance of policies. Agricultural policies in Switzerland support a heterogeneous farming community in terms of size and agro-ecological conditions and the evaluation tries to reflect this heterogeneity. The report also sheds light on the policy reform steps and decision making process in Switzerland and competiveness of food industries.

The study is structured along five Chapters: Chapter 1 – Assessment and recommendations; Chapter 2 provides an overview of the agricultural situation in Switzerland and the contextual information concerning the environment in which the agriculture sector operates and agricultural policies are implemented; Chapter 3 describes the agricultural policy reforms implemented since the mid-nineties and analyses the developments in the level and composition of support to agriculture; Chapter 4 assesses the impact of Swiss agricultural policy reforms on the economic and environmental performance of agriculture; and Chapter 5 evaluates strength and weaknesses of the Swiss food industries and their competitiveness on domestic and EU markets.

Acknowledgements

This report was prepared by the Trade and Agriculture Directorate of the OECD with the active participation of Swiss authorities and experts. The following people from the OECD Secretariat contributed to drafting this report: Václav Vojtech (project leader), Shingo Kimura, Jussi Lankoski, Sylvain Rousset and Frank van Tongeren. Statistical assistance was provided by Karine Souvanheuane, Alexandra de Matos Nunes and Véronique de Saint-Martin. Administrative and editing services were provided by Martina Abderrahmane, Jane Korinek and Michèle Patterson. Julien Hardelin provided expertise on the use of agri-environmental indicators, other useful comments were provided by Carmel Cahill and Andrzej Kwiecinski.

The Swiss Federal Office of Agriculture (OFAG) participated actively in providing data and feedback to the OECD Secretariat. Bernard Lehman (Director) and Jacques Chavaz (Deputy-Director) provided support in the launch of the process of the preparation of the study and in the round table discussion of the first draft of the study in Bern. Michael Hartmann (OFAG) provided invaluable effort to coordinate between the Secretariat the OFAG and the Swiss research institutions involved in the preparation of this report. Maurizio Cerratti, from the Swiss Permanent Delegation to OECD provided also valuable support to liaise with the Swiss authorities.

A number of experts from various research institutes and other institutions in Switzerland were involved by providing data and expert feedback to the Secretariat during the preparation of the study:

- Christine Bosshard, Daniel Bretscher, Felix Herzog, Pierrick Jan, Philippe Jeanneret, Markus Lips, Gabriele Mack, Stefan Mann, Thomas Nemecek, and Andreas Roesch (Agroscope Reckenholz Tänikon);
- Antoine Champetier de Ribes, Michel Dumondel, and Simon Peter (Agecon Group, ETH Zurich);
- Dominique Barjolle, Christian Schader, and Matthias Stolze (FiBL-Research Institute of Organic Agriculture);
- Christine Badertscher and Beat Röösli (Swiss Farmers Union – Union Suisse des Paysans);
- Florian Kohler and Franz Murbach (Swiss Federal Statistical Office).

Table of contents

Tables

Figures

Acronyms and abbreviations

AEI	Agri-environmental Indicators
AEM	Agri-environmental Monitoring
AP	Acronym for agricultural policy reform programmes
AP2002	Agricultural policy reforms applied in 1999-2003
AP2007	Agricultural policy reforms applied in 2004-07
AP2011	Agricultural policy reforms applied in 2008-13
AWU	Annual Working Unit
CAP	Common Agricultural Policy (of the European Union)
CH_4	Methane
CHF	Swiss franc
CO_2	Carbon dioxide
CPI	Consumer Price Index
DP	Direct Payments
EFTA	European Free Trade Association
ECA	Ecological Compensation Area
EU	European Union
EUR	Euro
FAO	Food and Agriculture Organization of the United Nations
FTA	Free Trade Agreement
GAO	Gross Agricultural Output
GATT	General Agreement on Tariffs and Trade
GDP	Gross Domestic Product
GHG	Greenhouse Gases
GNB	Gross Nitrogen Balance
GPB	Gross Phosphorus Balance
GSP	Generalised System of Preferences
GWP	Global Warming Potential
IMF	International Monetary Fund
IP LAIT	The Swiss dairy inter-branch organization
ISS	Institute for Sustainability Sciences
LDC	Least Developed Countries
LEI	Institute for Agricultural Economics (Netherlands)
LFA	Less Favoured Areas
MFN	Most Favoured Nation
N	Nitrogen
NACE	Statistical Classification of Economic Activities in the European Community
N_2O	Nitrous oxide
OECD	Organisation for Economic Co-operation and Development

OFAG	Swiss Federal Office for Agriculture (Office fédéral de l'agriculture)
OFS	Swiss Federal Statistical Office (Office fédéral des Statistiques)
P	Phosphorus
PEM	(Agricultural) Policy Evaluation Model
PEP	Proof of Ecological Performance
PPP	Purchasing Power Parity
R&D	Research and Development
RCA	Revealed Comparative Advantage index (or Balassa index)
RDP	Rural Development Plan
RMA	Relative Import Advantage index
RP 93-98	(Agricultural policy) Reform period 1993-98
RP 99-03	(Agricultural policy) Reform period 1999-2003 (see also AP2002)
RP 04-07	(Agricultural policy) Reform period 2004-07 (see also AP2007)
RP 08-12	(Agricultural policy) Reform period 2008-12 (see also AP2011)
RTA	Relative Trade Advantage index
RXA	Relative Export Advantage index
SAF	Society regrouping the only two sugar refineries in Switzerland
SMP	Skimmed milk powder
SNB	Swiss National Bank
SPS	Sanitary and Phytosanitary
SSG	Special Safeguard
TRQ	Tariff Rate Quota
UAA	Utilised Agricultural Area
UN	United Nations
UNFCCC	United Nations Framework Convention on Climate Change
URAA	Uruguay Round Agreement on Agriculture
USA	United States of America
USD	United States dollar
VAT	Value Added Tax
WTO	World Trade Organization

OECD indicators of support

ACT	All Commodity Transfers
CSE	Consumer Support Estimate
GCT	Group Commodity Transfers
GCT1	Group Commodity Transfers to All Crops
GCT3	Group Commodity Transfers to All Grains
GCT7	Group Commodity Transfers to All Livestock
GCT8	Group Commodity Transfers to Ruminants
GCT10	Group Commodity Transfers to grain and oilseeds
GCT11	Group Commodity Transfers to All Crops except wine
GSSE	General Services Support Estimate
MPS	Market Price Support
NAC	Nominal Assistance Coefficient
NPC	Nominal Protection Coefficient
OTP	Other Transfers to Producers
PSE	Producer Support Estimate
SCT	Single Commodity Transfers
TSE	Total Support Estimate

Executive summary

Agriculture plays a relatively minor and declining role in the Swiss economy, its share in gross domestic product is below 1% and the share in employment is around 4%. But the sector is perceived as an important element in maintaining food security, and as a provider of positive externalities such as environmental benefits and maintenance of cultural landscapes, which are highly valued by Swiss society. Much, but not all, of Swiss farming occurs in difficult natural circumstances.

Agricultural policies in Switzerland seek to find a balanced solution for addressing a variety of commercial, social and environmental objectives. The result is a system of market protection in combination with an elaborate set of payments to farmers that provides income support as well as incentives to certain types of farming practices.

The cost of agricultural policies is relatively high for Swiss consumers and taxpayers and currently represents about 1% of GDP. Current agriculture policies hold back further trade opening and hamper growth and export opportunities, in particular for the agro-food industry. Hence, reform of agricultural policies and related support to agriculture are important issues on the Swiss policy agenda.

The policy reforms implemented since the early 1990s substantially reduced market distortions. Domestic prices fell and moved closer to world market levels. Still, prices paid to producers are currently around 40% above the world market level. Although the level of farm support in Switzerland, as measured by the Producer Support Estimate (PSE), has declined gradually it is still one of the highest amongst OECD countries. In the mid-1990s around 70% of Swiss gross farm receipts came from public transfers either from consumers or taxpayers, this share was around 50% in 2011-13.

The gradual move away from policies supporting prices led to other ways to channel support to the farm sector, mainly through various payments afforded to farmers based on land or animal numbers. Re-instrumentation of policies improved the overall efficiency of delivering support to farmers because a larger share of support delivered through direct payments is captured by producers as compared to delivering it through supporting market prices. The re-instrumentation also allowed support to be more targeted to geographically less favoured areas.

Specific ecological payments are provided on a voluntary basis to farms that apply certain farming practices related to improving environmental performance and animal welfare; these represent less than 10% of the total payments. Switzerland has been amongst the forerunners in introducing environmental cross-compliance conditions, which make direct payments conditional on certain ecological requirements since 1999.

Significant progress has been made in meeting agri-environmental targets formulated in 2002 by the Federal government. By 2005 almost all targets were met except for the reduction of nitrogen surplus. The shift from price support to direct payments has reduced the intensity of inorganic fertilizer and pesticide use. Through providing incentives for an extensification of crop production and a shift from arable land to grassland, particularly in the plain region, policy reforms have led to positive environmental impacts at both the intensive (input use) and the extensive (land use) margin. Relative to the OECD average nitrogen surplus per hectare of agricultural land is slightly higher (8%) while for phosphorus it is significantly lower (50%).

A new agriculture policy framework is in place for the period 2014-17. The main policy change is the suppression of general area payments and improved targeting by tying payments to specific agricultural practices. Another important shift is the replacement of general headage payments to ruminants by an area payment to pastures with a requirement of a minimum stocking density. This will provide an incentive for further extensification of livestock production and may lead to lower stocking density.

To improve the performance of agricultural policies a useful distinction can be made between policies that address market failures (the provision of positive externalities and public goods as well as the avoidance of negative externalities), and those that address income problems. The current policies combine both aspects and seek to address markets failures by a combination of cross-compliance conditionalities and differential payment rates to stimulate certain farming practices and continuation of production in mountain regions.

Direct payments have now reached such a high level relative to what farmers earn by selling their products on the market that price and market signals appear to play only a secondary role in guiding their decisions. This hampers structural adjustment in the farm sector and, more generally, limits the development of a competitive food producing sector that contributes to food security objectives and continues to produce high quality products.

Benchmarking of the Swiss food and beverage industry against its main competitors in the EU, shows that competitiveness of the Swiss food and beverage industry is almost entirely driven by sub-branches that source most of their raw material inputs abroad or where inputs are non-agricultural (mineral water). The turnover of cocoa and chocolate manufacturing grew annually by 10%, almost twice as fast as the overall food and beverages industry (5.8%) in the period 2001-11. Together with beverages this industry counts for 72% of the exports of the Swiss agro-food industry.

The weakest sectors are meat and dairy processing, which mainly rely on domestic primary agriculture for their inputs. These industries as well as the weak animal feed sector, have to pay relatively high prices for their material inputs, well above the EU price levels. Additionally, these less competitive sectors have a relatively low growth of labour productivity and are relatively labour intensive.

Trade, including in agro-food products, is increasingly organized in global and regional value chains where specialized businesses at each stage of production add value to the product before it enters the final consumer market. Successful participation in such value chains requires unencumbered access to the best inputs at the lowest prices as well as regulations and technical standards that facilitate exchange of semi-processed and finished products with partner countries.

The development of a more market oriented commercial farming sector would contribute to increasing the competitiveness of those Swiss food industries that are mainly based on domestic agricultural raw materials. Lowering input costs while maintaining and enhancing the 'Swiss brand' image for domestic and foreign customers is likely to be a more sustainable strategy than shielding the industry from competitive forces. Structural change in the food and agricultural industry will continue, and will require exploiting scale economies and identifying niche markets.

The positive experience with opening the Swiss cheese market to EU competition since 2007 and eliminating the milk quota in 2009 shows that the farm sector has the capacity to adapt to market opening. A hypothetical policy simulation in this study shows that the gains to consumers from further aligning Swiss and EU agricultural prices would exceed the losses to producers and tax payers, even if complementary transitional payments are introduced. The impact on domestic production would be moderate overall, except for the beef sector. Important indirect positive effects on the food processing industry can be expected through lower prices for inputs and access to a larger consumer market.

The findings in this study lead to the following policy recommendations:

- The system of border protection needs to be further liberalised and trade barriers reduced. Export subsidies for processed products should be abolished.
- Reduce the overall level of general direct payments to allow farmers to respond to market signals and to increase incentives to produce high quality products at competitive prices.
- Introduce a two track system to reconcile the potentially conflicting objectives of the Swiss agricultural policy:
  - Under the first track, a differentiated direct payment system will secure the provision of goods and services to meet societal demand for e.g. cultural landscape, biodiversity;
  - In the second track potentially competitive producers (mainly in the plain region) should be allowed more freedom to optimise their production and respond to market signals. This second track may include policies facilitating structural change (investment support, exit strategies, etc.).
- Implement the two-track system by offering a regionally differentiated policy menu. The access to parts of the menu would be determined by the geographic location of producers, e.g. only mountain farmers get access to payments for cultural landscape services, while farmers in plain regions get access to support to modernize their business. This would not increase the administrative burden as the current system already uses geographical differentiation for direct payments.
- The role of regulations should be enhanced, and that of payments reduced to address objectives such as sustainable use of resources and animal welfare.
- Incorporate current cross-compliance requirements into mandatory regulation, which then provides the new baseline for new and more stringent cross-compliance requirements linked to support payments. This would reduce the budgetary burden and improve environmental performance of farming.

Chapter 1

Assessment and recommendations

This chapter draws conclusions concerning the reforms using the principles and operational characteristics established by Ministers for the evaluation of reform efforts across the OECD. It also where appropriate, makes recommendations concerning the continuing search for policy measures that are effective and minimally distorting, while contributing to the achievements of the objectives set by the society to the agricultural sector.

The context of agricultural policy reforms

Switzerland is a small open economy with a high GDP per capita and relatively low inflation and unemployment. Agriculture plays a relatively small role in the Swiss economy, accounting for less than 1% of GDP and around 4% of jobs. This reflects the highly developed industrial and services sectors in the economy. The farm structure is dominated by relatively small family farms. Hills and mountain farming areas are used for extensive milk and meat production. Arable land and irrigated land represent respectively 27% and 2% of total agricultural area. Switzerland has consistently been a net agro-food importer; the share of agro-food imports in total imports is around 6%, while the share of agro-food exports in total exports is around 4%.

Although agriculture plays a relatively minor and declining role in the Swiss economy, the sector is perceived as an important element in maintaining food security, and increasingly as a provider of positive externalities such as environmental benefits and maintenance of cultural landscapes, which are highly valued by Swiss society. The cost of Swiss agricultural policies is relatively high for consumers and taxpayers and currently represents about 1% of GDP. Hence, agricultural policies and related support to agriculture figure prominently in Swiss political debate.

Next to strongly voiced domestic societal demands, agricultural policy in Switzerland is also influenced by external drivers. The most important are:

- ***The WTO agreements,*** in particular the Uruguay Round Agreement on Agriculture (URAA), which liberalised agro-food trade and set a binding framework of disciplines on support to agriculture. In relation to its URAA commitments Switzerland has reduced its border protection which nevertheless remains significant due to the high starting level from which tariff reductions were made. On the other hand, the URAA commitments led to a restructuring of farm support provided through direct payments and a move towards less production and trade distorting forms of support.
- ***The gradual liberalisation of trade with the EU.*** The European Union is Switzerland's main trading partner, including for agro-food products. The agricultural agreement which came into force on 1 June 2002 facilitates mutual market access. The further liberalisation of the agro-food market with the EU (2005 and 2007) provided another incentive for further market oriented policy reforms in Switzerland especially in the dairy and sugar sector.

Assessment of policy developments

Development of support to agriculture 1986-2013

Although the level of support in Switzerland, as measured by the Producer Support Estimate (PSE), has declined gradually following the implementation of reforms started in the 1990s, it is still one of the highest amongst OECD countries. In the mid-1990s around 70% of Swiss gross farm receipts were from public transfers either from consumers or taxpayers. This share fell to around 50% in 2011-13. Switzerland is amongst the countries with the highest level of support as measured by the percentage PSE together with Japan, Korea, Norway and Iceland.

There has been a more marked improvement in the structure of support. The move away from policies supporting prices led to other ways to channel support to the farm sector, mainly through payments, which are considered less production and trade distorting. Due to a significant reduction in Market Price Support from the high levels of the 1980s, the share of the most production and trade distorting policies has been gradually reduced from 89% of total support in 1986-88 to 69% in 1995-97 and to 41% in 2011-13.

In the direct payments, the two most important categories are historical payments not requiring production (general area payment) and current headage payments for ruminants (adding up payments under various general programmes and those for less favoured areas).

Re-instrumentation and targeting of policies

During the 1980s agricultural policy was mainly based on supporting farm incomes through administered producer prices which were designated to cover producer costs. By the end of the 1980s, this system which guaranteed farmers fixed prices and guaranteed markets for their products, had reached its limits. The cost of the policy for the taxpayer (public expenditure) and the consumer (high prices) was rising continuously and the negative environmental impact of agricultural production was becoming more obvious. In addition, efforts towards world trade liberalisation created increasing pressure to relax protectionist and highly trade distorting measures in agriculture.

The policy reforms implemented since early 1990s substantially reduced market distortions. This led to a reduction in market price support as domestic prices came down and moved closer to world market levels. However, a significant level of price distortions remains as the prices paid to producers are still around 40% above the world market level. This is a significant burden on Swiss consumers.

The reduction in market price support was to a large extent compensated by increased direct payments to farmers. As a result the overall level of support to farmers declined only moderately during the past two decades. Most of those payments were general direct payments based on land area or animal numbers. Re-instrumentation of policies from market price support to direct payments improved the overall efficiency of delivering support to farmers because a larger share of support delivered through direct payments is captured by producers as compared to delivering it through supporting market prices. As a result, the producer surplus did not decrease by as much as the volume of support.

However, direct payments have now reached such a high level relative to what farmers earn by selling their products on the market that price and market signals appear to play only a secondary role in guiding their decisions. This may be hampering the development of a competitive food producing sector that contributes to food security objectives and produces high quality products. Moreover, the current size and structure of farm support is one factor contributing to the low competitiveness in Swiss food industries that rely on domestically sourced inputs. It also hampers the necessary structural changes in the farming sector by maintaining production where it is not economically viable and, more importantly, by restricting expansion and business development in the more productive plain regions.

Despite the increase in support targeted to geographically less favoured areas and efforts to improve environmental performance, some inconsistencies remain between policy instruments with different policy objectives. For example, the large increase in headage payments to maintain cattle production in geographically less favoured areas created incentives to increase stocking densities on grassland (pastures). This, in turn, has increased the environmental pressure from livestock farming.

Meeting agri-environmental targets

Agriculture plays an important role in the national sustainable development strategy. In addition to environmental regulation, *ecological payments* are provided on a voluntary basis to farms that apply certain farming practices related to improving environmental performance and animal welfare. While these payments are relatively low their share in total payments is rising. Switzerland has been amongst the forerunners introducing environmental cross-compliance conditions, with direct payments conditional on meeting certain ecological requirements that exceed mandatory requirements. Proof of Ecological Performance system (PEP) has been implemented since 1999.

The main environmental challenges facing agriculture were identified in 2002 by the Federal government which established a number of intermediate agri-environmental targets for 2005. Significant progress has been made in meeting those targets. In relation to the early 1990s almost all targets were met by 2005 except for the reduction in the nitrogen surplus. Developments of key agri-environmental indicators from 1990 to 2010 show that the most significant improvements in environmental performance already took place during the period 1990-92 to 1997-98, and since then the pace has slowed. The shift from price support to direct payments has reduced the intensity of inorganic fertilizer and pesticide use, and as a result it has reduced environmental pressures related to production intensity. Through providing incentives for extensification of crop production and a shift from arable land to grassland, particularly in the plain region, policy reforms have led to positive environmental impacts at both the intensive (input use) and extensive (land use) margins. According to impact evaluation studies, the requirements of environmental cross-compliance had a positive effect on farmland biodiversity, while contributing to the reduction of nitrate leaching and phosphorus pollution of surface waters.

While reforms in the 1990s have also reduced nutrient surplus and greenhouse gas emissions, subsequent later reforms have reversed the trend, driven by a policy-induced expansion in the livestock sector.

Despite overall improvements in environmental performance of Swiss agriculture some environmental challenges remain, including surface and groundwater water pollution from nutrients and pesticides.

Assessment of the AP 2014-17

The main element of the current policy change refers to the direct payment system. All general area payments have been eliminated. Direct payments to farmers are now closely related to certain objectives which are now in turn tied to agricultural practices. This is complemented by a system of transition payments to ease adjustment. Another important shift is the replacement of general headage payments to ruminants by an area payment to pasture while maintaining a minimal level of livestock husbandry.

The shift from headage payments to area payments will provide an incentive for an extensification of livestock production and may lead to lower stocking density. The reform is expected to reduce nutrient surplus and greenhouse gas emissions. The improvement of environmental performance is likely to be concentrated in hilly and mountain regions. The policy change towards payments based on area should also improve the transfer efficiency of the policy set. The simulation results show that producer surplus decreases only marginally, but the cost to taxpayers falls by much more.

Competitiveness of the Swiss food industry

The Swiss food and beverage industry overall has a relatively strong position in comparison to the main competitors in the EU. However, behind this aggregate observation lie vastly different positions among different sub-branches of the industry. In particular, the positive picture overall is mainly driven by one strong sub-branch "other food" manufacturing – dominated by cocoa and chocolate manufacturing. The turnover in this sub-branch grew annually by 10%, almost twice as fast as the overall food and beverages industry (5.8%) in the period 2001-11.

The two strongest sub-sectors "other food" and beverages industry count for 72% of exports of the Swiss agro-food industry. In these sectors a major part of the raw materials is imported or non-agricultural (mineral water).

The weakest sectors are meat and dairy, which rely mainly on domestic primary agriculture for their inputs, although some dairy producers successfully serve high value niche markets. These industries as well as the weak animal feed sector, have to pay relatively high prices for their material inputs, well above the EU price levels. Additionally, these less competitive sectors have experienced relatively low (increase in) labour productivity and are relatively labour intensive.

Further integration of the Swiss agro-food market with the EU market may provide the necessary impetus for structural changes in the less competitive sub-branches and will also strengthen their competitiveness *via* access to cheaper agricultural inputs. Competiveness of the agro-food industry can be enhanced by more transparent and less regulated markets both upstream and downstream and also food consumers would benefit from more competition in the downstream sector, including the retail level.

Future developments of policies – recommendations

Agricultural policies in Switzerland seek to find a balanced solution for addressing a variety of commercial, social and environmental objectives. The result is a system of market protection in combination with a set of payments to farmers that provides income support as well as incentives to certain types of farming practices. Also the regional (rural development) and sectoral (agriculture) policies can be better coordinated and the agricultural policy should be positioned in a broader context of rural policy (OECD, 2011).

A more explicit disentangling of policy objectives and instruments would improve the performance of the agriculture sector and enhance the efficiency of delivery of support to farmers. Business development, innovation and competitiveness of the farm sector and the food industry are hindered by trade policies that raise the prices for imported inputs and shield producers from competition.

Trade, including in agro-food products, is increasingly organized in global and regional value chains where specialized businesses at each stage of production add value to the product before it enters the final consumer market (OECD 2013). Successful participation in such value chains requires unencumbered access to the best inputs at the lowest prices as well as regulations and technical standards that facilitate exchange of semi-processed and finished products with partner countries. In this perspective, the system of border protection needs to be further liberalised and trade barriers reduced. Export subsidies for processed products should be abolished. Further market integration with the European Union could be a step in the near term. A policy simulation focusing on the effects on the primary agriculture sector shows that the gains to consumers from further aligning Swiss and EU prices would exceed the losses to producers and tax payers, even when some complementary transitional payments are introduced. The impact on domestic production would be moderate overall, except for the beef sector. Important indirect positive effects on the food processing industry can be expected through lower prices for inputs and access to a larger consumer market.

Security of food supply should be sought through a more competitive agriculture. Much, but not all, of Swiss farming occurs in difficult natural circumstances and support policies maintain production where it would not otherwise occur. A better distinction could be made, though, between policies that address market failures (the provision of positive externalities and public goods as well as the avoidance of negative externalities), and those that address income problems (OECD 2008). The current policies combine both aspects and seek to address markets failures by a combination of cross-compliance conditionalities and differential payments rates to stimulate certain farming practices and continuation of production in mountain regions.

Even farmers that would be competitive under more open market conditions cannot respond effectively to market signals because support payments have become too big a component in their receipts. Reducing these would be important to allow farmers to respond to market signals and to increase incentives to produce high quality products at competitive prices.

To reconcile the potentially conflicting objectives of Swiss agricultural policy, a differentiated policy approach should be considered, next to further liberalisation of agro-food markets. A two track system could be envisaged:

- Under the first track, a differentiated direct payment system will secure the provision of goods and services to meet societal demand for e.g. cultural landscape, biodiversity;
- In the second track potentially competitive producers (mainly in the plain region) should be allowed more freedom to optimise their production and respond to market signals. This second track may include policies facilitating structural change (investment support, exit strategies, *etc.*).

In practical terms, such a two-track system can be implemented by offering a regionally differentiated policy menu. The access to parts of the menu would be determined by the geographic location of producers, e.g. only mountain farmers get access to payments for cultural landscape services, while farmers in plain regions get access to support to modernize their business. This would not increase the administrative burden as the current system already uses geographical differentiation for direct payments. Persisting problems of low farm household incomes could be addressed through the social security system.

For some objectives such as sustainable use of resources and animal welfare the existing regulations could be made more stringent, while animal welfare and environmental compensation payments can be reduced. In practical terms current cross-compliance requirements can be incorporated into mandatory regulation, which then provides a new baseline for more stringent cross-compliance requirements linked to support payments. This could be achieved without increasing the regulatory burden to farmers or policy related transaction costs.

The development of a more market oriented commercial farming sector would also contribute to increasing the competitiveness of those Swiss food industries that are mainly based on domestic agricultural raw materials. Lowering input costs while maintaining and enhancing the 'Swiss brand' image for domestic and foreign customers are likely to be more sustainable strategies than shielding the industry from competitive forces. Structural change in the food industry will be inevitable, and will include exploiting scale economies and identifying niche markets.

References

OECD (2008), *Agricultural Policy design and implementation. A synthesis,* OECD Publishing, Paris, *http://dx.doi.org/10.1787/243786286663.*

OECD (2011), *OECD Territorial Reviews: Switzerland 2011*, OECD Publishing, Paris, *http://dx.doi.org/10.1787/9789264092723-en.*

OECD (2013), *Interconnected Economies: Benefiting from Global Value Chains*, OECD Publishing, Paris, *http://dx.doi.org/10.1787/9789264189560-en.*

Chapter 2

Agricultural policies in Switzerland: The policy context

This chapter provides an overview of the agricultural situation in Switzerland and the contextual information concerning the environment in which the agriculture sector operates and agricultural policies are implemented. This part focuses in particular on the role of agriculture in the economy, structural characteristics, and economic and environmental performance.

General aspects

Although agriculture plays a relatively minor and declining role in the Swiss economy, the sector is perceived as an important element in maintaining food security, and increasingly as a provider of positive externalities such as environmental benefits and animal welfare, which are highly valued by the Swiss society. Hence, agricultural policies and related support to agriculture are an important part of the Swiss political landscape. This chapter provides the contextual information on the economic, social, structural and environmental conditions which influence the Swiss agricultural sector and the climate in which agricultural policy is implemented.

Political and demographic characteristics

Switzerland is a relatively small country with 8 million inhabitants living on 40 000 km^2. It is located in the heart of Western Europe, at the intersection of German, French and Italian languages and cultures. Switzerland's four official languages, traditionally spoken in different regions of the country, are German, French, Italian and Romantsch (Rhaeto-Romanic).

Switzerland is a country with a very dynamic population growth when compared to other European countries. In the last decade the number of Swiss permanent residents increased by around 1% per year to reach 8.04 million in 2012. Since the 1990s immigration has accounted for most of the population increase. The share of immigrants in the total population was around 17% in 1990 and 23% in 2012. There have been no changes in the rural/urban population distribution as both in 1990 and 2012 the share of the rural population remains unchanged at 26% of the total.

There have also been no dramatic changes in the age structure of the population. Compared to 1990 the share of youth (0-19 years) declined by 3 percentage points to reach 20.4% in 2012; the share of population 20-64 remained relatively stable (62.2%), and the share of population over 65 years increased by 2.8 percentage points to reach 17.4%.

Politically and administratively Switzerland is a Confederation of 26 cantons (*Helvetic Confederation*). The Cantons (member states of the federation) enjoy a large degree of autonomy. Governments, parliaments and courts are organised on three levels – federal, cantonal, and communal).[1] Democracy and direct democracy in particular, have a long, but not undisputed, tradition in this country. Switzerland's unique political system is today one of the world's most stable democratic systems, offering maximal participation to its citizens.

The two chambers of Switzerland's national parliament meet several times annually, for sessions normally lasting three weeks. Member of Parliament is not a full time job in Switzerland, contrary to most other countries today. Members of Parliament in general continue to practise their profession to earn an income – thereby they are considered to be more cognizant of the concerns of their electorate.

Direct democracy applied through a system of referendums gives weight to ordinary citizens in their participation in the political process. All citizens may propose changes to the constitution if they obtain a given number of supporters (100 000 out of about 3 500 000 voters). The federal parliament discusses all proposals and may propose alternatives. All citizens then vote in a referendum whether to accept the original initiative, the alternate parliamentary proposal or to leave the constitution unchanged.

Geographical situation, natural resources and climatic conditions

Switzerland is one of the most mountainous countries of Europe. More than 70% of its area is covered by the Alps, in the central and southern regions, and the Jura, in the northwest. They are naturally wooded, with many cleared areas used as alpine pastures. Between these two mountain systems lies the *Swiss Plateau* (*Mittelland*), a basin that stretches across much of central Switzerland. The *Plateau* is a region covering 30% of Swiss area with an average elevation of about 580 m above sea level. The plateau contains many lakes and rivers, as well as Switzerland's most fertile soils. Most of Switzerland's large towns and about three-quarters of the Swiss population are located in this region.

The land use statistics provide four broad designations: settlement and urban areas, agricultural areas, wooded areas and unproductive areas such as lakes and rivers, unproductive vegetation, rocks and screes, glaciers and perpetual snow. Settlement and urban areas, accounting for 8% of the surface area, represent the smallest designation and agricultural areas (including alpine pastures), with a share of 36%, the largest. Wooded and unproductive areas occupy 31% and 25% of the land area respectively (Figure 2.1).

Figure 2.1. **Switzerland: Land use 2004-09**

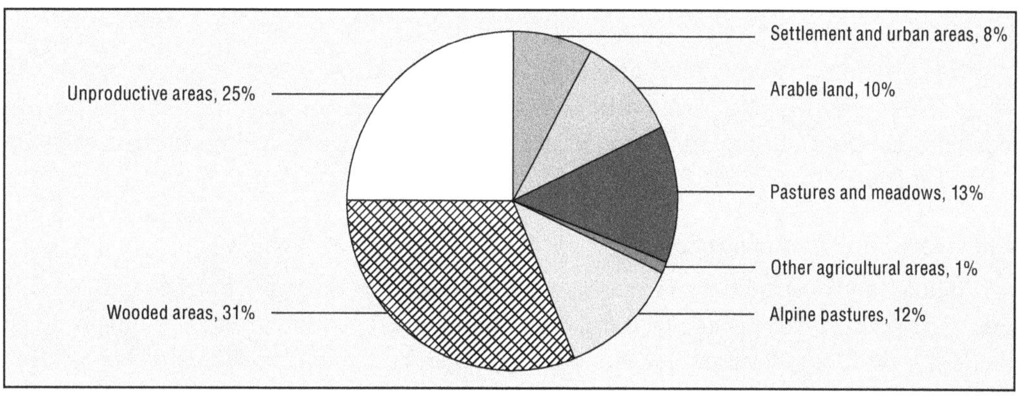

Source: Swiss Federal Statistical Office.

With its total surface area of 14 817 km^2, agricultural land represents the largest (36%) of the four major land use categories (Swiss Statistics, 2013). In 2009, meadows, farm pastures and alpine pastures accounted for two-thirds of the agricultural area, the remaining third being arable land and perennial crops. Due to the geographical configuration the agricultural land area is distributed unevenly across the country. The proportion of agricultural areas in the Central Plain region (49.5%) and the Jura (43.4%) is well above the national average. By contrast, a relatively small proportion of land is used for farming in the Western Central Alps (18.4%) and the Southern Flank of the Alps (12.7%).

Between 1985 and 2009, the total agricultural area shrank by 5.4% as a result of increases in settlement and urban areas, and wooded areas. Overall, more than half of the lost farmland was reused for settlement and urban development; the remaining portion went to wooded and unproductive land. New wooded areas primarily occupied abandoned alpine pastures situated at high altitudes (Swiss Statistics, 2013).

Switzerland is a principal **water** source in central Europe, and the nation's rivers flow into four different seas. Most rivers in Switzerland are not suited for navigation. Even the Rhine is not suited for commercial navigation in Switzerland until Basel, just inside the border with Germany. Lakes have long been important for transportation in Switzerland, and many towns are situated along lakeshores.

Waterpower is also a main natural resource of Switzerland. The principal source of water is runoff from the considerable annual precipitation that falls on the Alps. An important complement is melt water from the country's hundreds of glaciers. The Swiss have long harnessed the energy of falling water for productive uses. Today, the flow is captured by hundreds of hydroelectric power facilities, which provide 59% of the country's domestic electricity.

Switzerland has a varied **climate**, due largely to differences in elevation and exposure to sun and prevailing winds. On the plateau and in the lower valleys of Switzerland a temperate climate prevails, with a mean annual temperature of about 10° C. In the summer months, temperatures in low-lying areas can rise above 27° C, in winter months the temperature is generally below zero Mountainous areas are significantly cooler throughout the year, and temperatures decrease about 2° C for every additional 300 m of elevation. Large glaciers are found in the Alps, and permanent snow covers the highest peaks. Winter temperatures are generally below freezing throughout Switzerland, except for the north shore of Lake Geneva and the shores of the Swiss-Italian lakes, which have a mild climate like that of northern Italy.

Precipitation in Switzerland generally increases with elevation. Precipitation on the Swiss Plateau and in the lower valleys is about 910 mm annually; the higher regions typically receive more precipitation. Most precipitation occurs during the winter in the form of snow.

Macroeconomic performance

Switzerland is a small, open economy (as measured by the share of trade in GDP) with one of the highest GDP per capita. It is a developed market economy with a relatively stable macroeconomic environment. Moderate GDP growth is combined with relatively low levels of inflation and unemployment (Figure 2.2). The Swiss economy is mostly service oriented: the value added in the service sector represents 71% of GDP, while the manufacturing sector represents 28% and the primary sector, including agriculture is around 1%. Switzerland's main services sectors are banking and tourism.

Switzerland is one of the rare western European countries that have managed to grow over the past few years, thanks primarily to solid domestic demand. Household consumption growth has been supported by strong immigration, sustained consumer confidence and rising real wages (OECD, 2013a). Dynamic demography and historically low interest rates have boosted housing construction. Yet the unemployment rate has been edging up since mid-2011. Strong population growth, averaging around 1% per annum in recent years, has meant that in per capita terms, growth has been less impressive (OECD, 2013a).

Recently, the main economic challenge has been the strong appreciation of the Swiss franc, due to its safe haven status, which has been threatening to undermine competitiveness.

Figure 2.2. **Switzerland: Main macroeconomic indicators, 1990-2012**

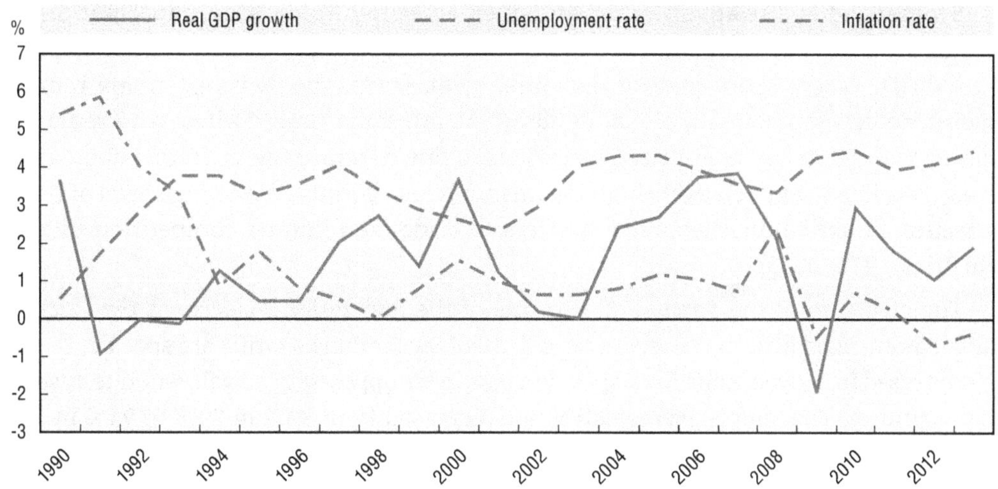

Source: OECD, Statistical databases, National Accounts, Labour Force statistics, Analytical Database, 2014.

In September 2011, the Swiss National Bank (SNB) decided to contain the appreciation of the Swiss franc (CHF) by setting a minimum exchange rate of CHF 1.20 per Euro and committing to unlimited purchases of foreign currencies when necessary (WTO, 2013).

In most recent years Swiss exports have been relatively sluggish by historical standards; nevertheless the current account surplus remains large, at 11% of GDP in 2012, driven primarily by financial services exports and investment income (OECD, 2013a). Monetary policy has been supportive since the beginning of the crisis, with near-zero policy rates since 2009. In addition, the minimum exchange rate has helped to contain deflationary shocks that cannot be addressed by further lowering the interest rate. Nevertheless, credit growth has been outpacing GDP expansion (OECD, 2013a).

One result of a new fiscal rule, the "debt brake rule", implemented by Switzerland since 2003 is that substantial fiscal surpluses were recorded at the federal level in years 2006-08. The surpluses allowed the implementation of economic stabilisation measures in 2009-10, mainly in the form of expenditure on roads, rail and other infrastructure investments and labour market measures, as well as a package of measures to mitigate the effects of the strong Swiss franc on the economy in 2011-12.

Fiscal policy is broadly neutral. A small general government surplus as well as continued economic growth should be sufficient to push gross government debt down further from its level of 44% of GDP in 2012. Public infrastructure, which will come increasingly under strain in the medium term due to sustained population growth and the transition to renewable energy sources, as well as education and R&D, might all be areas warranting public investment increases. Moreover, the budget will have to adjust to several structural pressures, including rising aging-related medical, disability and pension spending, and the broad array of existing and future subsidies including those foreshadowed in the government's climate change and nuclear phase-out strategy. Despite low income tax progressivity and modest cash transfers to households compared to other OECD countries, Switzerland enjoys a relatively equal distribution of disposable income, ranking around tenth most equal across the OECD. This results from a relatively flat wage distribution and very high rates of employment (OECD, 2013a).

Trade

Switzerland is a relatively small and open economy. Trade is an important driving force of the Swiss economy, with the share of trade in goods and services exceeding 100% of total GDP. Given the degree of dependence on trade, the Swiss economy remains exposed to developments in global demand. At the same time, Switzerland remains a country with relatively high price levels. This is due to a number of factors such as the strong Swiss franc and relatively high incomes, but also because of the high level of border protection in agriculture, technical barriers to trade, and limited competition in some industries (WTO, 2013).

The structure of the applied MFN tariffs of the Switzerland-Liechtenstein Customs Union has changed little in the recent period. All of Switzerland's tariffs are specific, i.e. they are expressed in a given value for a specific quantity as opposed to ad valorem rates used by most countries. The simple average MFN rate increased from 8.1% in 2008 to 9.2% in 2012, which reflects in part the appreciation of the Swiss franc. Tariff protection varies substantially across and within sectors, averaging 31.9% for agricultural products and 2.3% for non-agricultural goods according to WTO definitions (WTO, 2013). MFN duty-free treatment applies to almost 20% of all tariff lines, mainly products not produced domestically: fish, petrol, certain chemicals, base metals, etc. The preferential agreements concluded by Switzerland provide for free trade in most non-agricultural products, subject to certificates of origin. For agricultural products preferential access is provided mainly through bilateral tariff quotas. Least developed countries benefit from substantial and enhanced tariff preferences: all agricultural and non-agricultural products are duty and quota free. Switzerland's GSP rules of origin were harmonised with the EU's in 2011 (WTO, 2013).

Concerning the commodity structure of trade, the main imports in 2008-11 were machinery and transport equipment (around 27% of total imports), chemicals (21%), mining products (12%) and automotive products (7%). Agro-food products represented around 7% of total imports and were made up of mostly processed products. The structure of exports in the same period is dominated by chemicals (38%), machinery and transport equipment (21%) and other non-electrical machinery (11%). Agro-food products represented around 4% of total Swiss exports (again dominated by processed products). As far as the territorial structure is concerned, Swiss foreign trade is closely tied to the European market, in particular the EU, which accounts for almost 80% of its imports and around 60% of its exports. Germany is the most important country of origin for imports (representing 33% of Swiss imports) and export destination (20% of Swiss exports) within the EU. Other relatively important export destinations for Swiss goods and services are the United States (10%), China (4%) and Hong Kong, China (3%) (WTO, 2013).

Agricultural situation

Agriculture and the agro-food sector in the economy

The role of primary agriculture in the Swiss economy is minor and its share in the economy is shrinking due to the dynamic development of other sectors of the economy. The share of gross value added in agriculture in national GDP declined from 2.3% in 1990 to 0.7% in 2012. The share of agricultural employment in total employment also declined in this period from around 4.4% in 1990 to 3.5% in 2012. The high level of employment in agriculture compared to its contribution to GDP indicates a relatively low level of labour productivity compared with the other sectors of the economy, especially the services sector (Figure 2.3)

Figure 2.3. **Switzerland: Agriculture in the economy**

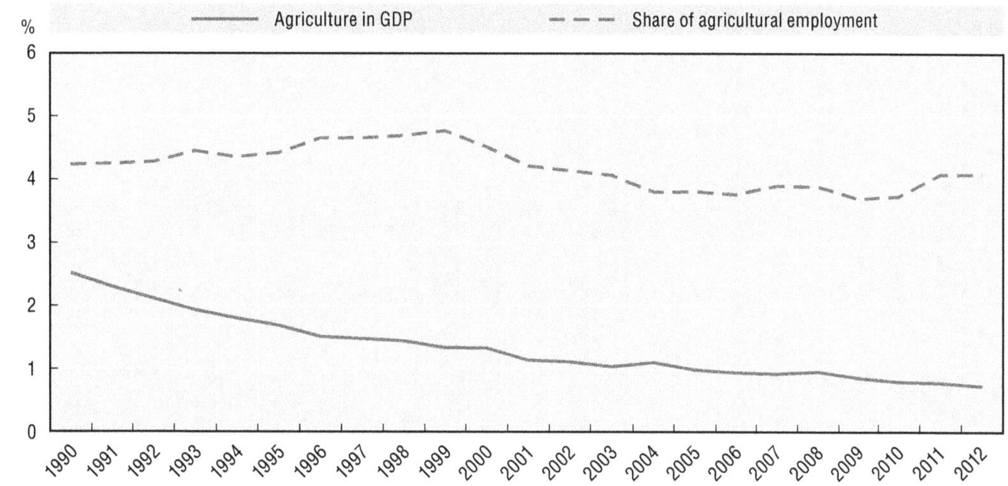

Source: OECD, Country statistical profiles, 2014 for the Agricultural value added; Labour Force statistics for the Employment by activity.

Overall, Switzerland has consistently been a net agro-food importer (Figure 2.4). The Swiss farming sector produces around 60% of the total domestic food consumption in calories (closer to 50% when imported feed is taken in account).

Figure 2.4. **Switzerland: Agro-food trade 1990-2012**

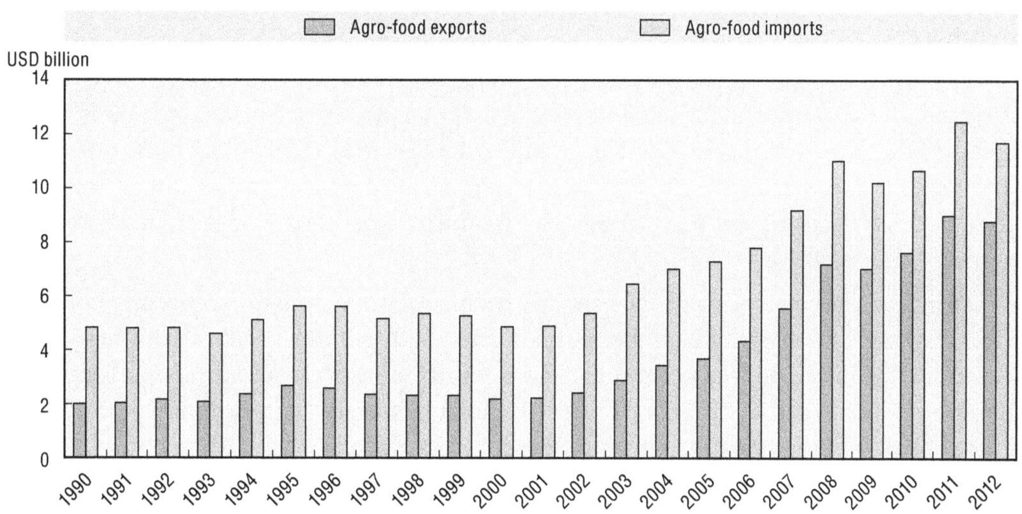

Source: OECD, International Trade by Commodity Statistics (ITCS) database, 2014.

Similarly, the share of agro-food trade in total trade is relatively small. Agro-food exports represented around 3% of total exports in 1990 and around 4% of total exports 2012. The shares of agro-food imports in the total declined from around 7% in 1990 to around 6% in 2012 (Figure 2.5). In both imports and exports the agro-food trade is dominated by processed food products while agricultural commodities play a minor role. Further insights into the Swiss agro-food trade are provided in Chapter 5 of this study, which examines the competitiveness of the Swiss agro-food industry.

Figure 2.5. **Switzerland: Share of agro-food trade in total trade, 1990-2012**

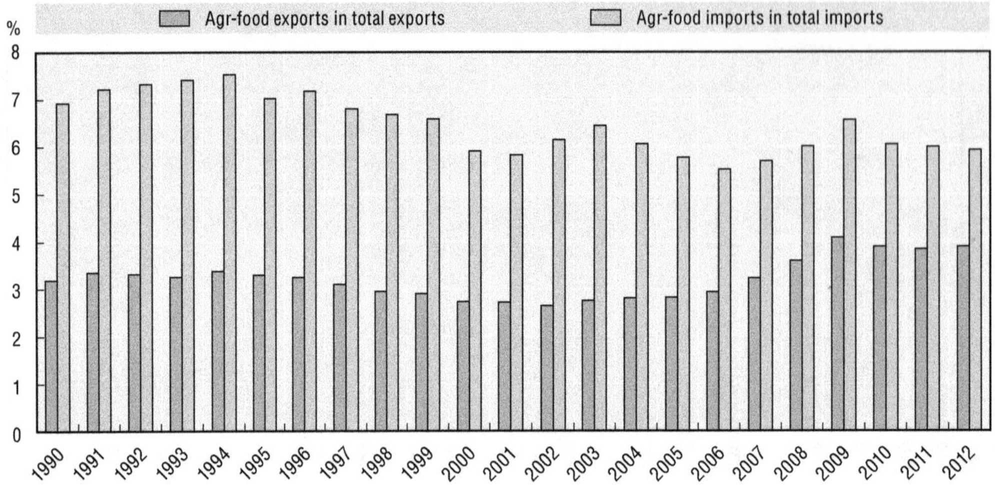

Source: OECD, International Trade by Commodity Statistics (ITCS) database, 2014.

Farm structure

The Swiss farming sector is made up mainly of relatively small family farms. In the past decades, Swiss agriculture underwent similar structural changes as other western European countries, i.e. moderate but continuous increase of the farm size and reduction of labour in agriculture. Although there has been almost no change in the agricultural output, the relative importance of agriculture in the national economy has been shrinking (see above).

In the period 2000-12, the number of farms declined from 70 537 to 56 575, a reduction of 1.8% per year. The pace of the decrease was slightly higher in the plains and mountain regions than in the hill region. The average farm area increased from 15.2 to 18.3 hectares. The largest reduction of farm numbers was among those that are from 3 to 10 hectares and from 10 to 20 hectares. On the other hand, the number of farm larger than 20 hectares increased (Figure 2.6).

Farms in the lowlands produce mainly arable crops (grains, oilseeds, silage maize, sugar beet and potatoes); most of the pig meat and poultry production is also located in the lowlands. The hilly and mountain areas are dominated by grassland and alpine pastures therefore the production of farms in these regions is mainly from ruminants (milk, beef and veal, and to a lesser extent sheep and goat production).

Farm employment

As in other developed countries, the use of labour in agriculture in Switzerland has decreased during the past decades, together with the restructuring of farms and increase in labour productivity. In the period 1990-2012, the total labour force in agriculture decreased from 253 000 to 162 000 (-36%). This reduction was greater as regards family labour which dropped by 40%, while non-family (hired) labour was reduced by 14% over the 1990-2102 period. The decrease in the agricultural labour force was also more pronounced for full-time farmers (-43%) than for part-time farmers (-29%). The number of foreign workers in agriculture was rather stable and its share in the total agricultural labour force increased from 5.6% in 1990 to 9% in 2012 (Figure 2.7).

Figure 2.6. **Share of agricultural area in farm size categories, 1996-2012**

Per cent (TAA = 100%)

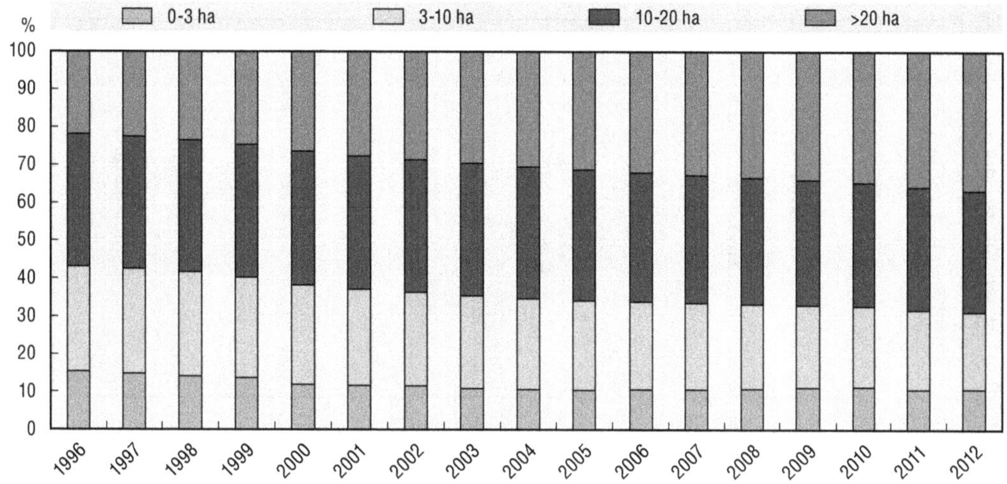

Source: Swiss Federal Statistical Office.

Figure 2.7. **Switzerland: Structure of labour force in agriculture (numbers)**

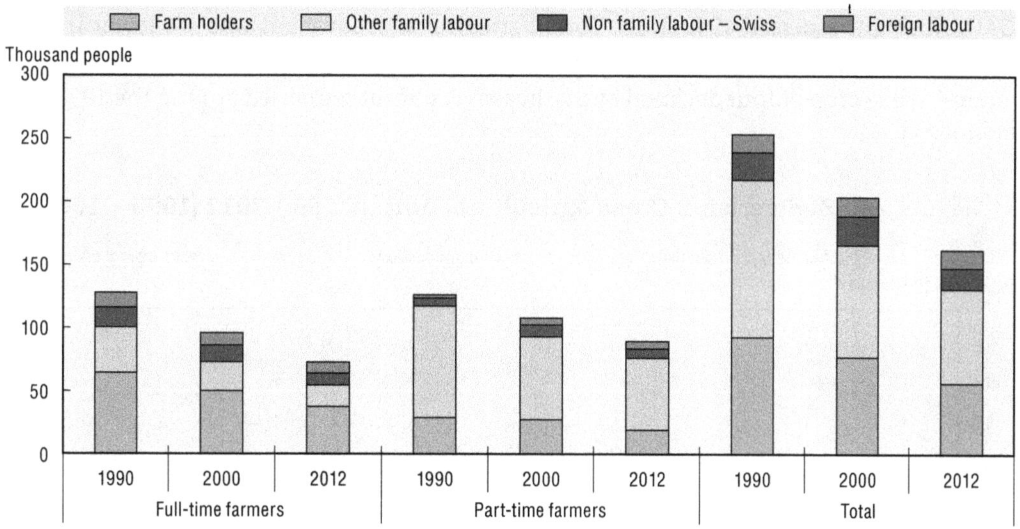

Source: Swiss Federal Statistical Office.

Agricultural employment fell by 38% from 1990 to 2012, from 127 000 full time equivalent workers to 79 000. This reduction broadly corresponds to a similar decrease in the labour force in the EU15 of 40% over the same time period.

The gender balance in the farm labour force was broadly constant over the last two decades. Men represented 63% and women 37% of the total agricultural labour force in 2012 (Figure 2.8). This is similar to the situation in 1990 (64% and 36% respectively). However, the gender disparity is even stronger when considering some types of farm employment. Ninety-five percent of farm holders were men in 2012 despite a small increase in the number of female farm holders.

Figure 2.8. **Switzerland: Gender structure of agricultural labour force (%)**

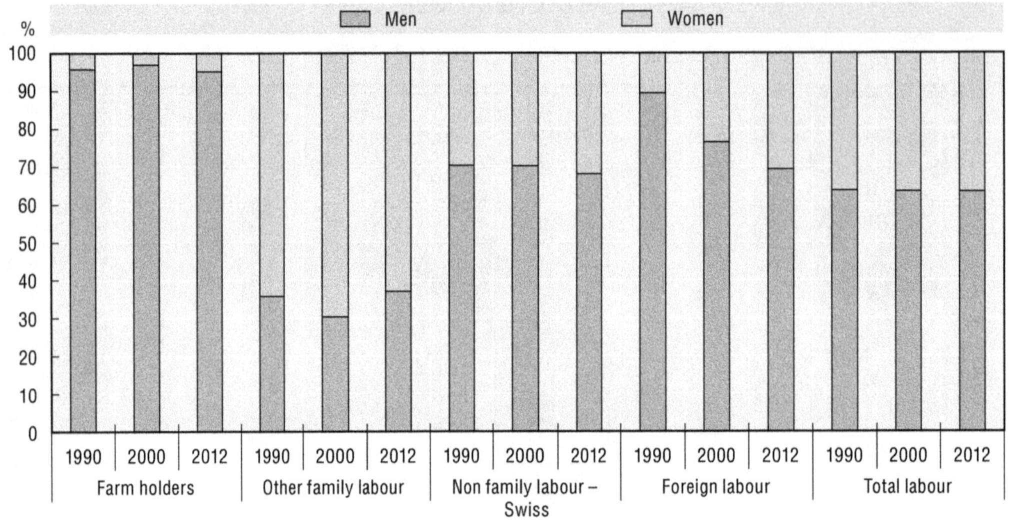

Source: Swiss Federal Statistical Office

Farm output

There were no dramatic changes in the overall volume of Gross Agricultural Output (GAO) during the period 1990-2012. Total agricultural production was 4% lower in 2012 compared with 1990. There were, however, some structural changes in agricultural output volume: while crop output declined by 9%, livestock output remained around the 1990 level (Figure 2.9).

Figure 2.9. **Switzerland: Gross agricultural output, 1990-2012 (1990 = 100)**

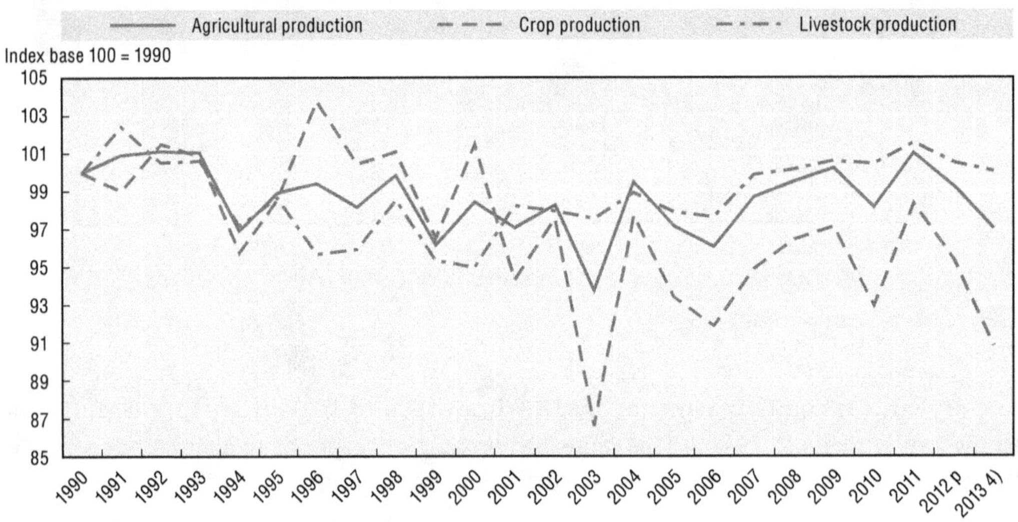

Source: Swiss Federal Statistical Office.

Figure 2.9 reflects the changes in physical output (constant 1990 prices), however the changes in the value of production (in real prices) were also affected by a marked decline for domestic prices for crop products (grains, oilseeds) and some livestock products. The changes in the volumes and prices resulted in important shifts in the structure of the value of production. In 2012, crop products represented about half the value of total agricultural production. The crops sector is dominated by fruits, vegetables and fodder crops. Vegetables and horticulture represent 16% of total agricultural production, fodder crops 12%, fruits and wine 11%. The main arable crops that figure prominently in some other OECD countries' agricultural production contribute a much lower share to total agricultural production in Switzerland: grains represent 4%, sugar beet 2% and oilseeds 1%. The most important livestock products are related to cattle rising: milk represents 23% of total agricultural production and beef 14%. The other livestock sectors represent a smaller share (pig meat 9%, and poultry and eggs 5%).

There were changes in the structure of the value of agricultural production during 1990-2012. The most dynamic change was for vegetables and horticulture, rising from 9% of total agricultural production in 1990 to 16% in 2012. In turn the share of grains dropped from 9% to 4%. In the livestock sector the shares were relatively stable. Exception is poultry production for which the share increased from 1% to 3% of total agricultural production.

In the crop sector the largest drop was for potatoes (66% of 1990 output) and grains (reduced by 29%), while the oilseeds and sugar beet output increased by 55% and 37% respectively over the same period. The output of feed crops remained around the 1990 level. The volume of vegetable production was up by 29% while the volume of fruit production went down by 18%.

Although total livestock output remained unchanged, there were some structural shifts. The poultry meat production has more than doubled, and egg production increased by 21%. Some other sectors increased more moderately: sheep output went up by 6% and milk production by 3%. Beef and pig meat output declined by 12% and 5% respectively.

Agriculture productivity

Switzerland has relatively little arable land. The utilised agricultural area (UAA) per inhabitant was 0.134 hectare in 2011 (10% lower than in 2000), and for arable land it was 0.034 ha/inhabitant (16% down). However, as intensive production is perceived as a nuisance to the environment, many agricultural policies support extensive farming practices. Organic production represents more than 10% of agricultural area. The use of fertilisers has declined and is relatively low compared with the EU (part 2.3 of this chapter gives more details).

The yields of main crops are relatively low compared with EU countries. Grains and oilseed yields were relatively stable, or even declined for some varieties, in 1995-2012. The yields of potatoes and sugar beet have increased, although potatoes are produced on a smaller area. The yields in the milk production, a key sector of Swiss agriculture, increased and compensated the reduction in milk cow numbers.

Land productivity has been stable over the last two decades. Gross agricultural output per hectare of agricultural land oscillated around CHF 10 000 (Figure 1.10). The ratio of labour to agricultural land declined from 10 Annual Working Units (AWUs) per 100 hectares in 1997, to 7.5 AWUs/100 hectares in 2012. Gross agricultural output per unit of labour has increased sharply since 1991 but has been relatively flat since 2007 (Figure 2.10).

The value of agricultural output produced by each unit of labour fell sharply over the last two decades. The value added in agriculture at real market prices per unit of labour declined, from around CHF 30 000/AWU in 1991 to CHF 20 000/AWU in 2012. This downward trend reflects the reduction in domestic prices paid to producers. In contrast,

Figure 2.10. **Switzerland: Land and labour productivity, 1997-2012**

GAO in CHF per ha of UAA and per AWU

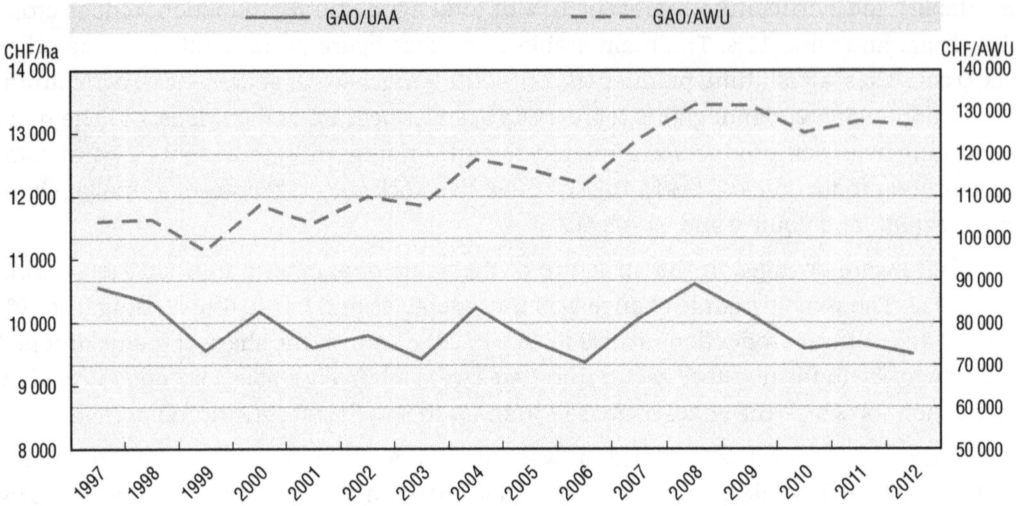

Note: GAO/UAA: Gross Agricultural Output in constant prices per hectare of Used Agricultural Area (excluding alpine pastures).

GAO/AWU: Gross Agricultural Output in constant prices per Annual Working Unit.

Source: Calculations based on Swiss Federal Statistical Office data.

the farm revenue at factor costs per AWU increased in the same period due to an increasing role of direct payments in farm revenues (from CHF 39 000 in 1991 to CHF 55 000 in 2012) (Figure 2.11).

Figure 2.11. **Switzerland: Labour productivity indicators, 1997-21-012**

Value added and farm revenue in CHF per AWU

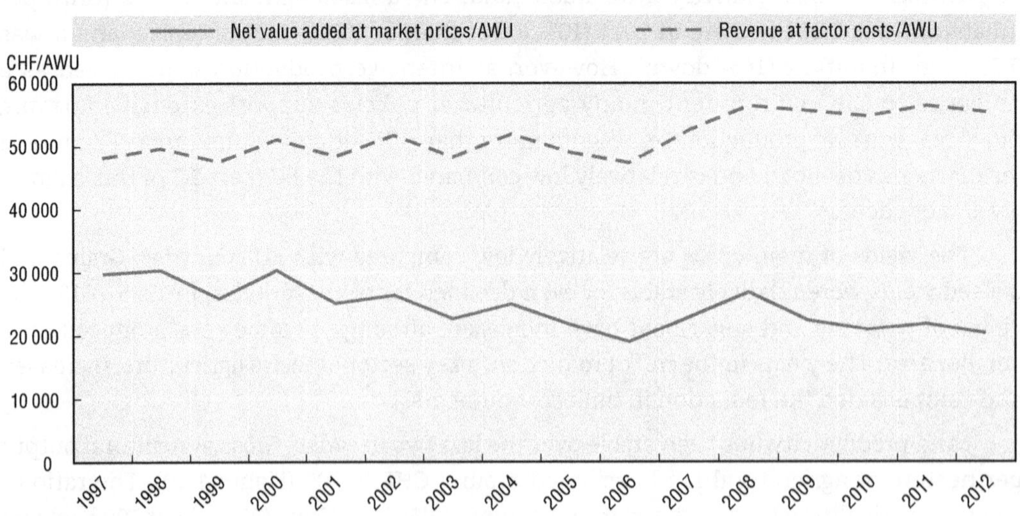

Source: Calculations based on Swiss Federal Statistical Office data.

Food consumption

Food prices paid by consumers in Switzerland are relatively high compared with other European countries. On average, the prices of food and non-alcoholic beverages in Switzerland were 53% above the EU27 in 2011. Higher price levels also apply to other goods and services. The prices for housing and energy, for example, are more than double those in EU27. Despite high price levels, Swiss households spend proportionally less on food and non-alcoholic beverages than households in the EU. Swiss expenditure on food and beverages, excluding restaurant meals, is around 7% compared with 10% in the early 1990s. This is comparable to the level in Germany.

Environmental performance of Swiss agriculture

Evolution of selected OECD agri-environmental indicators 1990-2010

Agriculture plays a key role in the national sustainable development strategy. The main environmental challenges facing agriculture were identified in 2002 by the Federal government which established a number of intermediate agri-environmental targets for 2005. Using a 1990-92 baseline, targets included reducing nitrogen surpluses by 23%, reducing phosphorus surpluses by 50%, lowering pesticide use by 32%, lowering emissions of ammonia by 9%, transferring 10% of farmland into ecological compensation areas. In addition, 98% of farmland was targeted to be cultivated according to ecological compliance or organic farming standards and 90% of drinking water in agricultural areas was required to have a nitrate level below 40mg/l (Badertscher, 2005; OFAG, 2004; Herzog and Richner, 2005; Flury, 2005).

Almost all agri-environmental targets have been met with the exception of nitrogen surplus for which the reduction was smaller than target (Table 2.1). From 1990-92 to 2006-08 inorganic fertiliser use fell by 21% for nitrogen fertiliser and 59% for phosphorus fertiliser, and pesticide use fell by almost 14%, albeit the latter has since increased. These important steps toward agri-environmental sustainability have however required trade-offs. During the same period, direct farm energy consumption increased by 46% (OECD, 2013b).

Table 2.1. **Quantitative agri-environmental policy targets in Switzerland**

	Baseline	Target 2005	Target achievement
Nitrogen balance	96 000 tons (1994)	74 000 tons (23% reduction)	18% reduction in 2005
Phosphorus balance	20 000 tons (1992-1994)	10 000 tons (50% reduction)	By 2002 reduction by 65%, but only 10-30% reduction of P-pollution of waters
Pesticides	2 200 tons of active ingredients (1990-92)	1 500 tons (32% reduction)	
Ammonia	57 300 tons (1990)	52 143 tons (9% reduction)	48 300 tons (16% reduction) by 2010
Biodiversity	Agricultural area 1.08 million ha	10% of total agricultural area as ecological compensation area of which 65 000 ha in the valley region	64 505 ha of ecological compensation in the valley area in 2012
Nitrate		90% of agricultural catchments below 40mg/litre	Goal achieved by 2002/03
Agricultural area	Agricultural area 1.08 million ha	98% of area under proof of ecological performance (PEP) or organic	97.8% in 2005

Source: Schader, 2009; Herzog et al., 2008; OFAG, 2012 and Kupper et al., 2013.

Agricultural nutrient surpluses have decreased by 17% for nitrogen and 72% for phosphorus over the period 1990-92 to 2006-08. The nitrogen surplus per hectare of agricultural land (68 kg/ha) is slightly higher than the OECD and EU15 averages for nitrogen, while in the case of phosphorus (3 kg/ha) it is considerably lower than the OECD average and the same as the EU15 average (2006-08) (Table 2.2).

Table 2.2. **Nitrogen and phosphorus balances in Switzerland, 1990-2009**

	Average (thousand tonnes or kg/ha)			Average annual % change	
	1990-92	1998-2000	2006-08	1990-92 to 1998-2000	1998-2000 to 2006-08
Nitrogen balance, thousand tonnes of N	122	99	102	-2.6	0.3
Nitrogen balance, kg per ha of agricultural land	80	65	68	-2.4	0.5
Phosphorus balance, thousand tonnes of P	17	6	5	-13.1	-1.7
Phosphorus balance, kg per ha of agricultural land	11	4	3	-12.2	-2.3

Source: OECD (2013), *OECD Compendium of Agri-environmental Indicators.*

Much of the reduction in nutrient surpluses is explained by lower fertiliser use, especially inorganic fertiliser (Figures 2.12 and 2.13). This is particularly significant in the case of phosphorus fertiliser and, to a lesser extent, greater use of livestock feeds containing less phosphorus (OFAG 2002). Nutrients in livestock manure were reduced by 4% for nitrogen and 7% for phosphorus; and crop uptake of nutrients showed only a small reduction over this period.

Figure 2.12. **Nitrogen balance in Switzerland since 1990**

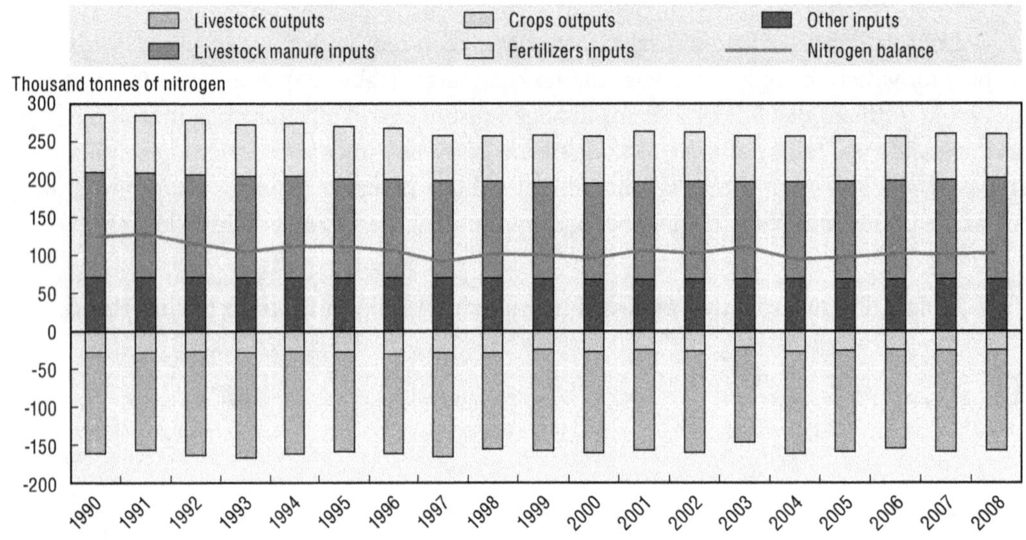

Source: OECD (2013), *OECD Compendium of Agri-environmental Indicators.*

The decrease in nutrient surpluses occurred for the most part in the 1990s. Most of the 18% decrease in nitrogen surpluses occurred during the period from 1990-92 to 1997-99. This decrease in nitrogen surplus can be primarily attributed to reduction in mineral fertilizer use and livestock manure. Since then, nitrogen surplus has increased by 4% from 2000-02 to 2006-08, largely explained by a rise in manure nitrogen inputs. Phosphorus surplus decreased from 1990-92 to 1997-99 by 65% while from 2000-02 to 2006-08 it decreased by 17%. Phosphorus surplus reduction is due to significant and continued decrease of phosphorus fertilizer use. The introduction of ecological direct payments and environmental cross-compliance (proof of ecological performance, PEP) and its requirement of balanced use of nutrients have contributed to decreased nutrient surpluses, especially during the first years after their implementation until most of the farms had joined these programmes.

Figure 2.13. **Phosphorus balance in Switzerland since 1990**

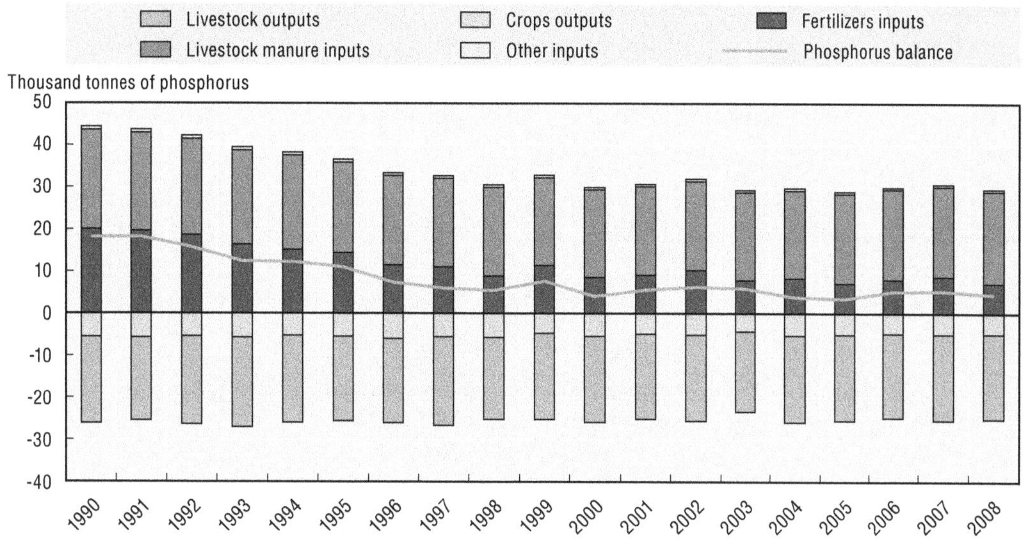

Source: OECD (2013), *OECD Compendium of Agri-environmental Indicators*.

Swiss farms have obtained better results from the nutrients they use. The efficiency of nitrogen use improved from 57% to 61% over the period 1990-92 to 2006-08, while for phosphorus it increased from 61% to 84%. This reflected a 59% fall in the use of inorganic phosphorus fertilisers while the crop uptake of phosphorus decreased by 12%. Moreover, most farms and farmland were under a nutrient management plan, with around 90% of farms in 2000-03 conducting soil nutrient tests (OFAG, 2005). In addition, manure storage capacity rose by over 50% from 1990 to 2003 (OFS, 2005).

Despite reductions in nutrient surpluses, water pollution from agricultural nutrients persists in arable farming regions (Swiss Agency for the Environment, Forests and Landscape, 2002; Badertscher, 2005; Herzog and Richner, 2005). Farming contributes around 40% of nitrates and over 20% of phosphorus in surface water. With respect to nitrates in groundwater, the share of agriculture is 75% (Swiss Agency for the Environment, Forests and Landscape, 2002). The concentrations of nitrates in groundwater in monitoring points in agricultural areas have declined from around 20mg/l in the mid-1990s to 18mg/l in 2003. Over 10% of monitoring points (risk areas) in arable cropping areas have nitrate concentrations greater than 40 mg/l (OFS, 2005). About 3% of monitoring points in agricultural areas exceed drinking water standards, although this share is low compared to many other OECD countries. The share of monitoring sites in agricultural areas that exceed recommended drinking water limits for nitrates in groundwater is 5%.

The situation regarding the use of pesticides over the two decades under review has been more uneven. Quantities of pesticides sold from 1990 to 1999 decreased by 33%, but increased by 41% from 2000 to 2010. However, there is a break in time series and thus these two periods are not comparable. About 62% of groundwater monitoring sites in agricultural areas showed the presence of one or more pesticides in 2009. In arable farming regions in 2010, 10% of groundwater monitoring sites had pesticide concentrations in excess of drinking water standards.

Despite a substantial 16% decrease in ammonia emissions from agriculture from 1990 to 2010, farming's share in total ammonia emissions remained high at 92%. Much of the decrease in ammonia emissions, which vary regionally, has resulted from improvements in livestock manure storage and manure application. The medium term target of AP14-17 is to reduce ammonia emissions to 41 000 tonnes per year and a long term target is 25 000 tonnes per year. Under the *Gothenburg Protocol* Switzerland agreed to reduce total ammonia emissions to 63 000 tonnes by 2010 and this target was met by 2009-10 with emissions down to 62 500 tonnes.

Agricultural emissions of greenhouse gases (GHGs), which contributed 11% of national GHGs (2008-10) declined by 7% between 1990-92 and 2008-10 (Figure 2.14). Methane (CH_4) emissions account for 56% of total CO_2 equivalent emissions; the share of nitrous oxide (N_2O) was 44% in 2010.

Figure 2.14. **Development of key agri-environmental indicators in Switzerland, 1990-2010**

Index 1990 = 100

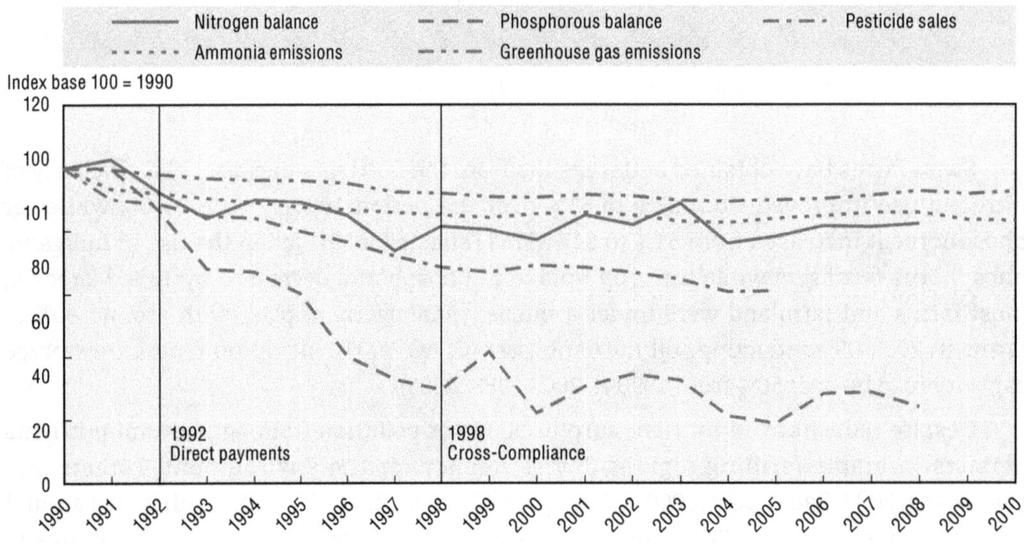

Source: OECD (2013), *OECD Compendium of Agri-environmental Indicators*.

Success in reducing emissions and GHG in agriculture has been contrasted with a strong increase in energy use. Direct on-farm energy consumption increased by 45% between 1990 and 2010, although it accounted for only 1.3% of total national energy consumption in the latter period. Direct consumption is comprised of, e.g., fuel and electricity while indirect consumption includes the energy used for manufacture of inputs, such as fertilizer and machinery. As regards direct energy consumption fuel consumption has increased at higher rate than electricity consumption. With regard to indirect consumption there has been increase in energy consumption related to machinery and imported fodder while energy consumed in relation to fertilisers has dropped (OFAG, 2011).

The growth in ecological compensation areas (ECAs) has eased the pressure from farming on biodiversity. The diversity of crop varieties and livestock breeds used in production has risen over the period 1990 to 2002 (OFAG, 2005). There are also programmes for conservation of crops and livestock *in situ* and extensive gene bank collections *ex situ*, while all endangered native livestock breeds are included under conservation programmes. A high share of the nation's flora and fauna uses farmland as primary habitat, including mammals (75%) and invertebrates (55%, butterflies, 40% of grasshoppers), although the share is lower for birds (22%). However, the share of endangered birds using farm habitat is 50%.

The area of agricultural semi-natural habitats under ECAs expanded from 2% to 13% of farmland from 1993 to 2012. Over 85% of the ECAs are extensive and low intensity managed meadows, and about 50% of ECAs are in lowland areas (OFAG 2005; Badertscher 2005). Evaluation studies show mixed results for the impact of ECAs on flora and fauna (Knop et al., 2006; Herzog et al., 2005). ECAs seem to have enhanced biodiversity in contrast to intensively managed farmland, although there are important variations between different ECA types (Knop et al., 2006; Herzog et al., 2005). Species abundance and richness of ECAs with meadow litter and hedgerow seem to be higher than ECAs for hay meadow and traditional orchard. This reflects the impact of intensive management practices (Herzog et al., 2005). The ecological quality of mountainous ECAs was significantly higher than that in lowlands (Herzog and Richner, 2005; Flury, 2005; Herzog et al., 2005).

Conversion of farmland to other uses has had adverse impacts on ecosystems and cultural landscapes. The fragmentation of agricultural land by urban and transport development, the conversion of farmland to mainly urban use, and the abandonment of farmland in marginal areas have had an adverse impact on farmed ecosystems and cultural landscapes (OFS, 2005). There has however been an increase in some linear landscape features on farmland, such as hedges and dry stone walls. ECAs are also reported to have reduced the effects of farm habitat fragmentation by serving to interconnect habitat sites (Badertscher, 2005).

In conclusion, areas under agri-environmental schemes have expanded and most environmental targets for agriculture have been met. Since the increase in expenditure on agri-environmental measures from the early 1990s, farmer participation in these schemes has grown to nearly 90% of all farms and 98% of farmland (OFAG, 2005).

Comparison of agri-environmental indicators for Switzerland, EU and OECD

Figure 2.15 provides comparison of key agri-environmental trends for OECD, EU15 and Switzerland over two decades. In Switzerland the sales of pesticides reduced at a faster rate than the change in crop production especially from 1990-92 to 1998-2000 while for OECD and EU15 the sales of pesticides were mainly reduced during the period from 1998-2000 to 2008-10. Nitrogen and phosphorus surplus reduction has taken place in all cases and in Switzerland most of the reduction happened from 1990-92 to 1998-2000 while for OECD and EU15 relatively larger reductions occurred from 1998-2000 to 2008-10. In all cases there has been relative decoupling of nutrient surpluses from agricultural production so that production has decreased relatively less than nutrient surpluses.

Brief summary of main developments as regards environmental performance of Swiss agriculture

Significant progress has been made in meeting the government's agri-environmental targets. In relation to the early 1990s base almost all targets were met by 2005 except nitrogen surplus reduction. Developments of key agri-environmental indicators from 1990 to 2010 show that environmental performance improvements largely took place already over the period 1990-92 to 1997-98, and since then pace of performance improvements has slowed down. Despite overall improvements in environmental performance of Swiss agriculture some environmental challenges remain including surface and groundwater water pollution from nutrients and pesticides.

Figure 2.15. **Key agri-environmental trends for OECD, EU15 and Switzerland from 1990-92 to 2008-10**

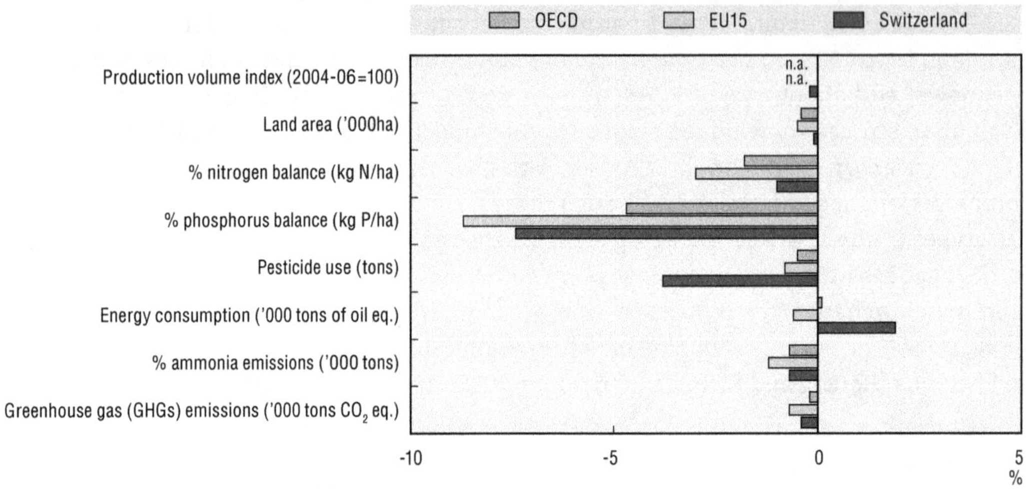

Note: n.a. – not available

1. For Switzerland, agricultural pesticide sales refer to the period 1990-92 to 1998-2000.

Source: OECD (2013), *OECD Compendium of Agri-environmental Indicators*.

Box 2.1. **Agri-environmental Monitoring**

In Switzerland the Federal Office for Agriculture (FOAG) is carrying out agro-environmental monitoring (AEM) based on the Federal Act on Agriculture (Art. 185) and the Ordinance on the Evaluation of Sustainability in Agriculture. The goal of the AEM is to assess the impact of agriculture on the environment. A set of seventeen agro-environmental indicators (AEI) provides the base for the AEM. The AEI are split in six topics (nitrogen, phosphorus, energy/climate, water, soil and biodiversity) and two types of indicators (driving forces and environmental effects). As the competence centre for AEI, the Institute for Sustainability Sciences (ISS) at Agroscope is responsible for the centralized evaluation of AEI, including the development of AEI methods. Data for the calculation of the AEI are collected since 2009 in a network of currently 300 farms to obtain agro-environmental information at the regional level and by the type of farm.

"Species and habitats of agriculture" is an indicator programme delivering information about the state and dynamics of biodiversity in agricultural landscapes of Switzerland and therefore works at the landscape level. To assess the state and dynamics of species diversity in agricultural landscapes, four groups of indicators have been developed, 35 in total: 1) diversity of habitats and structures, 2) quality of habitats and structures, 3) species diversity and 4) species quality. An additional group of indicators covering diversity and quality of ecological compensation areas has been added as an integrated evaluation of policy measures within the monitoring programme.

Notes

1. Small communes have citizens' meetings instead of parliaments; local courts are shared among several communities.

References

Badertscher, R. (2005), "Evaluation of Agri-environmental Measures in Switzerland", in OECD, *Evaluating Agri-environmental Policies: Design, Practice and Results*, OECD Publishing, Paris, *http://dx.doi.org/10.1787/9789264010116-en*.

Flury, C. (2005), *Evaluation des mesures écologiques et des programmes de garde des animaux*, Swiss Federal Office for Agriculture, Berne, Switzerland, *www.blw.admin.ch/imperia/md/content/evaluationen/050920_agrokol_tierwohl_f.pdf?PHPSESSID=ef9470b4" target="webpage">www.blw.admin.ch/imperia/md/content/evaluationen/050920_agrokol_tierwohl_f.pdf?PHPSESSID=ef9470b4*.

Herzog, F. and W. Richner (eds.) (2005), *Évaluation des mesures écologiques : Domaines de l'azote et du phosphore*, Les cahiers de la FAL 57, Institut de recherche en écologie et agriculture, Zurich-Reckenholz, Switzerland, *www.reckenholz.ch/" target="webpage">www.reckenholz.ch/*.

Herzog, F., S. Dreier, G. Hofer, C. Marfurt, B. Schüpbach, M. Spiess and T. Walter (2005), "Effect of ecological compensation areas on floristic and breeding bird diversity in Swiss agricultural landscapes", *Agriculture, Ecosystems and Environment*, Vol. 108, pp. 189-204.

Herzog, F., V. Prasuhn, E. Spiess, and W. Richner (2008), "Environmental cross-compliance mitigates nitrogen and phosphorus pollution from Swiss agriculture", *Environmental Science and Policy*, Vol. 2, pp. 655-668.

Knop, E, D. Kleijn, F. Herzog and B. Schmid (2006), "Effectiveness of the Swiss agri-environmental scheme in promoting biodiversity", *Journal of Applied Ecology*, Vol. 43, pp. 120-127.

Kupper, T., C. Bonjour, B. Achermann, B. Rihm, F. Zaucker and H. Menzi (2013), "Ammonia emissions for Switzerland 1990 to 2010 and previsions until 2020", Report in German with English and French Summary.

OECD (2013a), *OECD Economic Surveys: Switzerland 2013*, OECD Publishing, Paris, *http://dx.doi.org/10.1787/eco_surveys-che-2013-en*.

OECD (2013b), *OECD Compendium of Agri-environmental Indicators*, OECD Publishing, Paris, *http://dx.doi.org/10.1787/9789264186217-en*.

OECD (1990), *National Policies and Agricultural Trade: Switzerland*, OECD Publishing, Paris.

OFAG (2012), Office federal de l'agriculture, *Rapport Agricole 2012*, Swiss Federal Office for Agriculture, Berne. Switzerland, *www.blw.admin.ch/" target="webpage">www.blw.admin.ch/*.

OFAG (2011), Office federal de l'agriculture, *Rapport Agricole 2011*, Swiss Federal Office for Agriculture, Berne. Switzerland, *www.blw.admin.ch/" target="webpage">www.blw.admin.ch/*.

OFAG (2005), Office fédéral de l'agriculture, *Rapport Agricole 2005*, English summary Agricultural Report 2005, Swiss Federal Office for Agriculture, Berne. Switzerland, *www.blw.admin.ch/"target="webpage">www.blw.admin.ch/*.

OFAG (2004), Office fédéral de l'agriculture (2004), *Rapport Agricole 2004*, English summary Agricultural Report 2005, Swiss Federal Office for Agriculture, Berne. Switzerland.

OFAG (2002), Office federal de l'agriculture, *Rapport Agricole 2002*, English summary Agricultural Report 2005, Swiss Federal Office for Agriculture, Berne. Switzerland, *www.blw.admin.ch/" target="webpage">www.blw.admin.ch/*.

OFS (2005), Office fédéral de la statistique, *Agriculture in Switzerland 2005*, Swiss Federal Office for Statistics, Berne, Switzerland.

Schader, C. (2009), "Cost-effectiveness of organic farming for achieving environmental policy targets in Switzerland", Ph.D. thesis, Institute of Biological, Environmental and Rural Sciences, Aberystwyth, Aberystwyth University, Wales. Research Institute of Organic Agriculture (FiBL), Frick, Switzerland.

Swiss Agency for the Environment, Forests and Landscape (2002), *Environment Switzerland 2002*, Berne, Switzerland, *www.umwelt-schweiz.ch/buwal/eng/publikationen/index.html" target="webpage">www.umwelt-schweiz.ch/buwal/eng/publikationen/index.html*.

WTO (2013), World Trade Organisation, *Trade policy review: Switzerland and Lichtenstein*, Geneva, Switzerland.

Chapter 3

Policy trends and support to agriculture in Switzerland

This chapter describes the agricultural policy reforms implemented since the mid-nineties. The chapter outlines the driving principles of the policy reforms including motivations and changes in priorities, and discus the reform process such as sequencing and consensus building. The chapter also analyses the developments in the level and composition of support to agriculture resulting from the agricultural policies implemented within the analysed period. This analysis relies mainly on the PSE/CSE/GSSE and related indicators.

Agricultural policy framework

Agricultural policy objectives

The Swiss agricultural policies implemented in the mid-1990s reflect the societal consensus to have an agricultural sector which meets market demands and operates in an environmentally friendly way, together with providing public goods to the society (e.g. biodiversity and cultural landscape). Also ethical aspects of animal welfare are an important concern raised by Swiss society. Moreover, there is an increasing effort to coordinate the agricultural policy with regional (Cantonal) rural development policies.

Since the mid-1990s the main objectives of Swiss agriculture are qualified in an article of the Constitution. In a popular vote held in 1996, a clear majority (more than three-quarters) of the participating electorate voted in favour of adding a corresponding article on agriculture to the federal constitution. The tasks for agriculture set out in this article include the following main objectives:

- To make a major contribution towards ensuring food supplies for the population, although Switzerland is and will remain a major food importer.
- Ecological standards are an important objective of agricultural policy; production methods will ensure that future generations will have fertile soil and clean drinking water.
- Taking care of the landscape is seen as an essential task of agriculture. A varied landscape is viewed to contribute to the quality of life for the population, and is at the same time the basis for a flourishing tourist industry which is an important asset of the national economy.
- Finally, encouraging decentralised settlement to help maintain rural areas is an aim which is also supported by regional policies.

Although these policy objectives are clearly defined in principle, policy makers face the challenge of implementing them at a lower cost to the society, reconciling their potentially contradictory (or unwanted) effects.

Main driving forces of the agricultural policy reforms

The political decision making process

The specific features of the Swiss political system have an impact on the direction and pace of agricultural policy reforms. The relatively large autonomy of the 26 cantons in the Swiss Confederation, in combination with elements of direct democracy mean that political decision making processes involve many stakeholders and extensive consultations. Popular initiative is one of the key elements in the Swiss referendum system. Ordinary citizens may propose changes to the constitution or other legislative acts, if they can find a number of supporters (100 000 out of about 3 500 000 voters, smaller numbers at the cantonal and communal level) in a non-mandatory referendum. The parliament discusses the proposals, which most often leads to an alternative proposal, and all citizens may decide subsequently in a referendum whether to accept the original initiative, the alternate parliamentary proposal or to leave the constitution or legislation unchanged.

In general, the agricultural policy framework is set for a four year period. In case of delays in the process of agreement of the new policy set, the initial period of application is extended. From the early 1990s, the various policy reforms were driven through the following periods: 1993-98 (Low on Agriculture article 31a and 31b); 1999-2003 (AP2002); 2004-07 (AP2007); and 2008-13 (AP2011). A policy package for the next Agricultural policies was agreed in 2013 and is scheduled to be implemented for the period 2014-17.

As for most political decisions, the decision making process around agricultural policies includes elements of direct democracy. As a consequence, the system of implementation of policy changes is a rather lengthy but well-structured procedure, which provides an opportunity to participate in the decision making process to all stakeholders and representatives of various elements of the society (even individuals). This all-embracing decision making process helps to build consensus around the measures proposed and minimises the risk that the proposed legislations may be rejected in a non-mandatory referendum.

In general the process of the preparation and adoption of the new federal legislation (*Ordonnance du Conseil Fédéral*) related to the implementation of a policy reform consist of the following steps:

- The Federal Council outlines the main characteristics of the proposed policy reform including drafts of the various legislative acts (drafting and planning phase – *phase d'élaboration*) and submits it in a broad consultation process.
- Cantons and all concerned organisations, and even individual citizens may express their views in these consultations.
- In a next step, a revised draft is consolidated by the Federal Council, including final legislative and budget proposals, and transmitted as *Message du Conseil Fédéral* to the Parliament.
- The Parliament (*Conseil des États* = Council of States or Senate, and *Conseil national* = National Council or Chamber of Representatives) debates the document including any proposed amendments. Do both councils agree on both the Federal Council's dispatch and proposed amendments after maximal three rounds of debates (if necessary including a conciliation conference), the proposed policy reform and the new federal legislation will enter into force.
- At this final stage, there is still a possibility for the opponents of this legislation to call for a non-mandatory referendum concerning the debated legislation (see information above on conditions for such referendums). If conditions are met, the referendum takes place. If the legislation is approved it is implemented, if not the proposal is rejected. The same accounts if there is no agreement after the parliament conciliation conference.
- An additional cycle starts after final approval of the legislative amendments. The Federal council elaborates draft proposals for implementation rules of the new policy ("ordinances") and submit them to the Cantons and all concerned organizations.
- Taking the comments from the consultation process into account, the Federal Council consolidates and approves the final version of the implementation rules, including corresponding budget allocations, and put them into force.

This process may at least partly explain the fact that the process of reforming the Swiss agricultural policy that started in the 1990s is rather gradual and implemented in several phases. As these policy reforms were built on a general societal consensus and includes lengthy debates with all stakeholders, the reaction to the implementation of these policies is generally positive (even if various stakeholders often have conflicting interest in terms of the impacts of the policies applied).

External (international) drivers for policy change

Switzerland is a relatively small and open economy with important ties to the world markets. In terms of the agro-food trade it is mainly tied to the EU market both for imports and exports. Hence the decisions on agricultural policy changes in Switzerland are also driven by external drivers. The most important are:

- The WTO agreements, in particular the Uruguay Round Agreement on Agriculture (URAA), which liberalised agro-food trade and set a binding framework on support to agriculture. Agriculture in Switzerland is greatly affected by efforts being made to further deregulate world trade. In relation to its URAA commitments Switzerland has reduced its border protection which nevertheless remains important due to a relatively high level of negotiated base for tariff reduction. On the other hand, the URAA commitments led to a restructuration of the support provided through direct payments and a move towards less production and trade distorting forms. In WTO negotiations, Switzerland, along with some other countries, proposes that more attention be paid to non-trade concerns related to the multifunctionality of agriculture.
- Bilateral trade agreements and especially the gradual liberalisation of trade with the EU. The European Union is Switzerland's main trading partner for agro-food products. The agricultural agreement which came into force on 1 June 2002 facilitates mutual access to markets. The further liberalisation of the agro-food market with the EU (2005 and 2007) provided another incentive to further market oriented policy reforms especially in the dairy and sugar sector.

Agricultural policy developments

Policies as of early 1990s

At the end of the 1980s and in the early 1990s, Swiss agriculture was isolated from world market signals by important trade barriers and heavy domestic market regulations. On average the domestic prices paid to farmers were 4.5 times higher than world market prices. In addition most payments were based on output (mainly market price support) or input use, and in the form of area and headage payments related to specific products. Support for farmers was high, as almost 80% of gross farm receipts were from transfers from agricultural policies. Also around 80% of farm support was provided through policies which are potentially most production and trade distorting.

Most of the support was provided in the form of Market Price Support (MPS) which reflects the price gap between domestic and world prices resulting from the high tariff barriers but especially heavy interventions on domestic market. Figure 3.1 provides an illustration of the price gap for the various agricultural products for the period of 1986-88, as measured by the Producer Nominal Protection Coefficient (NPC).

At the end of the 1980s, agricultural policy, which guaranteed farmers fixed prices and guaranteed markets for their products, had reached its limits. The cost of such a policy for

Figure 3.1. **Switzerland: Producer Nominal Protection Coefficient by commodities**

Coefficient, average 1986-88

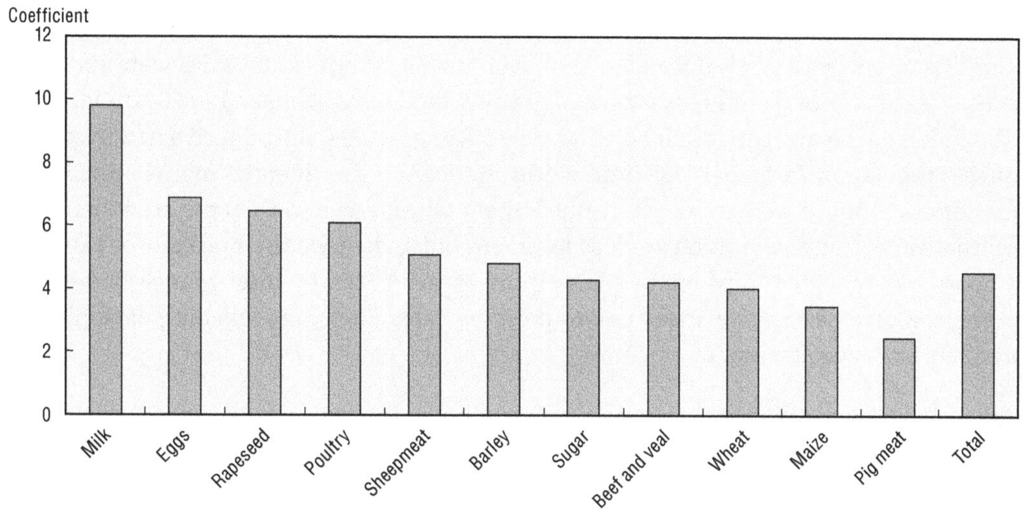

Source: OECD (2014), "Producer and Consumer Support Estimates", *OECD Agriculture statistics* (database).

the taxpayer (public expenditure) and the consumer (high prices) was rising continually and the negative environmental impact of agricultural production was becoming more obvious. In addition, efforts to liberalise trade created increasing pressure to reform protectionist and trade distorting measures in agriculture. Time was ripe for reforming the Swiss agricultural policy system.

Policy reforms during the 1990s

Following a decision taken by the Parliament, substantial reforms of agricultural policy were implemented from 1993 to 1998 (Article 31a and 31b) and another reform step was taken in 1999-2003 (AP2002). There were three main elements of the agricultural policy reform package applied during the 1990s:

- More transparent import regime, reduction of tariff barriers and gradual abolition of heavy domestic market intervention. During the 1990s, state guarantees for prices and markets were gradually removed, resulting in a reduction of prices paid to farmers. Maintaining their market share became an important challenge to Swiss farmers.
- Introduction of direct payments not tied to specific products as compensation for lower prices and as remuneration for the farmers' public and ecological services. The structure of these payments was further revised from 1999.
- The introduction of environmental cross-compliance was part of the reshuffle of the system of direct payments. Since 1999, all direct payments have been based on stringent proof of ecological performance (*Performances écologiques requises*).

Reduction of market support

Border measures: The system of quantitative restrictions was replaced by tariff quotas with a partial reduction of in-quota tariffs and a reduction of threshold prices used for calculations of import tariffs for feeding stuff such as grains, soybean meal, feeding oil, gluten feed. The latter had the main objective to reduce the feed costs for Swiss livestock producers. Although the system of border measures became more transparent and the level of protection reduced, the level of tariff barriers remained relatively high.

Domestic market regulation: Interventions on the domestic market have been substantially reduced. The heavy administrative tools guaranteeing prices and production quantities were gradually reduced. All state guarantees for prices and sales were abolished in 1999 (except price guarantees for bread grains which were abolished in 2001). Only milk production quotas were maintained and a new sugar act (1998) introduced a new system of market regulation for sugarbeet production. Hence, on the internal market prices and quantities produced were more determined by supply and demand. On the other hand, the reforms supported tools which are less interventionist vis-à-vis the market. For example, support was provided to promote self-help measures taken by inter-branch bodies and producers' organisations or measures to promote sales, including labelling of traditional products from specific areas.

New system of direct payments

Prior to the reforms initiated in the 1990s, most direct payments were provided for specific products in the form of area/headage payment (*acreage premiums* for wheat, coarse grains, potatoes; *Suckle cow premium, etc.*), and in the form of input subsidies (*feed grain price reduction*). There were also payments (mostly for ruminants) provided to farmers in the mountain areas (*holding of livestock in mountain areas, summer pasturing*). However, the total amount of these payments was relatively low compared to transfers provided through market price support in the period prior to the 1990 reforms (Figure 3.2).

Figure 3.2. **Market price support and direct payments in the total Producer Support Estimate**

Million CHF

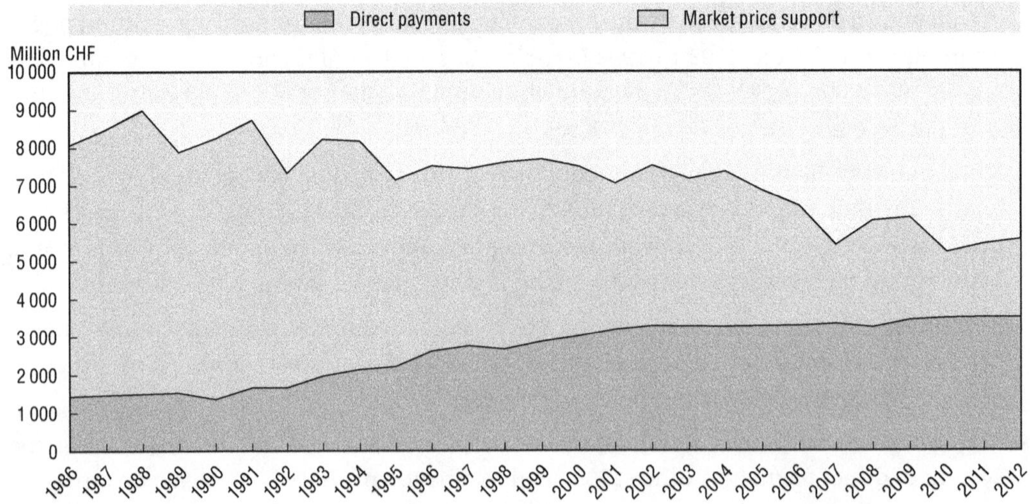

Source: OECD (2014), "Producer and Consumer Support Estimates", *OECD Agriculture statistics* (database).

A new system of ***direct payments*** has been gradually introduced starting from 1993. Some of the existing product specific payments were phased out: acreage premium for wheat and potatoes (discontinued in 1989); suckler cow premium (1998); acreage premium for coarse grains and feed grain price reduction (2000). But other product specific payments were introduced to compensate for the price reduction related to market liberalisation:

milk price supplement for cheese production (introduced in 1996) and *payment for oilseed cultivation* (2000).

However the most important change in the system of direct payments was the introduction of two main broad categories of new payments in 1993: i) *General direct payments* and ii) *Ecological Direct payments.*

Newly introduced ***General direct payments*** are non-commodity specific payments. They were implemented in two distinct periods: 1993-98, and from 1999, and represent by far the most important part in the total direct payments (Figure 3.3).

Figure 3.3. **Structure of direct payments, 1986-2012**

Million CHF

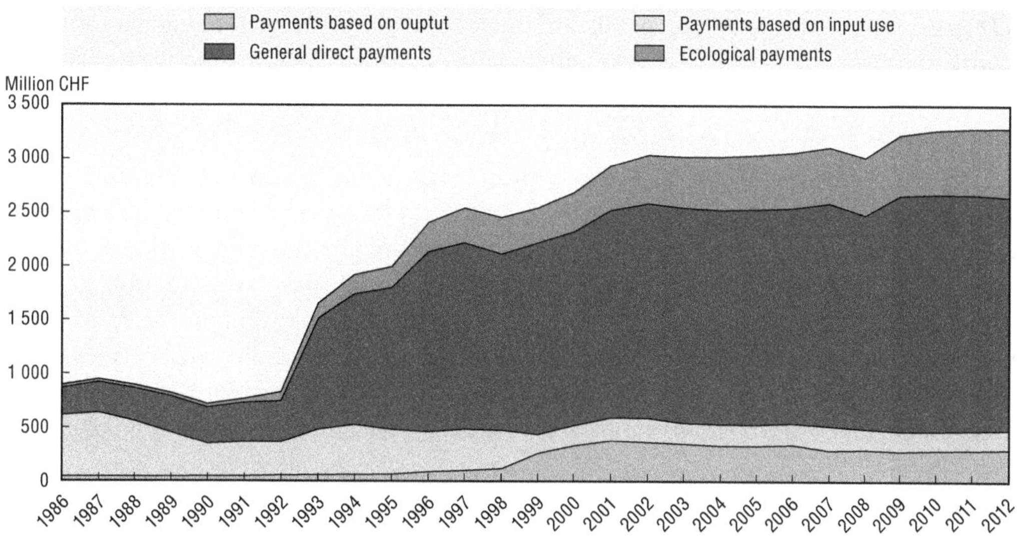

Source: OECD (2014), "Producer and Consumer Support Estimates", *OECD Agriculture statistics* (database).

From 1993 to 1998, the General direct payments consisted of three main elements:

1. *Complementary Direct Payments*: payments based on various criteria consisting of four elements: i) base farm payment; ii) base area payment – arable land; iii) base area payment – grassland; and iv) supplementary payments. A level of the Complementary Direct Payment per individual farm was a result of the combination of these four elements.
2. *Payments for integrated production*: payment per hectare of crops cultivated according to specific integrated production standards (related to biodiversity, soil conservation, manure application, cropping plan, cultivar selection, integrated pest management, and the holding of livestock); at least 5% of the land is cultivated as extensive meadows or floral fallow. The rate of payment decreases by half above 50 hectares of cultivated land. A supplement is granted if the integrated production applies to the whole farm.
3. *Payments for farming in difficult conditions*: this part of general direct payments consists of programmes which were already implemented in the previous period i.e. holding of livestock in mountainous areas (headage payment) and farming on steep slopes (area payment).

From 1999, the system of the General direct payments was reorganised and was applied without major changes up to 2013. The main changes compared with the previous package are:

1. The system of *Complementary Direct Payments* was replaced by a general *Area Payment* – payment per hectare of agricultural land independent of any requirements to produce particular crops. The payment is subject to the income and asset ceilings for direct payments.
2. The *Payments for integrated production* were discontinued. On the other hand, the ecological requirements introduced in the system of environmental cross-compliance (in order to receive any of the general direct payments producers have to comply with these requirements) is largely based on the requirements for integrated production (see Box 3.1)
3. A general *Payment for ruminants* was introduced in 1999. It is a payment per animal (cow, horse, sheep, goat, etc.). The rate of payment varies between species, and is increased if animals are held for summer pasturing, and reduced if milk is marketed.
4. *Payments for farming in difficult conditions*: this part of general direct payments remained without major changes. A new payment for *wine cultivation on steep slopes* was introduced in 1999.

Ecological direct payments are another category of newly introduced direct payments during the 1990s reforms. These payments are designed to provide additional remuneration to farmers for providing non–marketed goods and services such as biodiversity, landscape , animal welfare, etc. Some of the programmes provide incentives for more sustainable use of resources and to reduce pollution. Participation in these programmes is voluntary. Overall, the sum of the ecological direct payments is much lower, compared to the general direct payments, but these payments recorded a continuous upward trend in the period 1993-2013, which reflects increasing participation of farmers in those programmes and a larger choice of programmes becoming available (Figure 3.3).

Ecological compensations consist of programmes providing remuneration for ecological services such as creating valuable habitats for animals and plants and cultural landscape. Various programmes were gradually implemented under this heading:

- From 1993, payments for extensive meadows; less intensively used meadows for forage production; floral fallow; and payments for tall fruit trees (with trunk and crown);
- In 1999, additional programmes were added within the Ecological compensation: payments for extensive dry land and litter area; hedges and rustic groves; rotation of fallow land; and extensive area strips.

Besides the Ecological compensations, there are other programmes in the Ecological direct payments:

- Payments for *summer pasturing* (already implemented in the pre-reform period) were included in this package. Farmers receive payment for transhumance on condition that they use the alpine pastures in an environmentally friendly way;
- *Extensive cereal and rapeseed farming* (from 1992) provides area payments to farmers complying with the criteria set for extensive modes of production;
- *Organic farming* of crops (from 1993): payment per hectare of special crops, open arable land excluding special crops, and green areas and litter meadows producing according to specific organic farming regulation requirements on the entire farm;
- *Animal welfare* programmes are also part of the Ecological direct payments. There are two main programmes: i) payments for *regularly keeping animals outdoors* (from 1993) is a payment per head of animal kept outdoors for a certain number of days every week/ month; and ii) *animal welfare through housing systems* (from 1996) is a payment per head of

animal raised in stabling systems according to specific standards (with a minimum of 5 livestock units). The rate of payment in these two programmes varies according to the animal species. The requirements for these programmes are more stringent than the terms of legislation on animal protection. In 2002, around 30% of all animals were kept in particularly animal-friendly conditions and 60% had regular outdoor exercise.

- *Payment for measures to protect water quality.* A separate programme aims specifically to improve the quality of water in problem areas.
- *Contribution to environmental quality* (from 2001) is an additional programme aimed at raising the quality of the ecological zones and encouraging farmers to link them up.

Environmental cross compliance: Since 1999, farmers receive direct payments only if they meet certain ecological requirements – Proof of Ecological Performance, PEP (*Prestations écologiques requises*, PER). The key elements of proof of ecological performance are outlined in Box 3.1.

Box 3.1. **Environmental cross-compliance (Proof of Ecological Performance, PEP)**

In order to be eligible for general direct payments Swiss farms need to fulfil the criteria of the proof of ecological performance (PEP). Thus, PEP represents environmental cross-compliance linked to agricultural support payments. PEP rules are defined in Article 70 of the Federal Law on Agriculture. The main PEP criteria are:

- Balanced nutrient use: maximum 10% surplus of nitrogen and phosphorus as shown by a farm's nutrient balance (based on crop requirements)
- Minimum share of ecological compensation areas (ECA): at least 7% of a farm's utilised agricultural area has to be allocated as ecological compensation area (e.g. extensive meadows, low intensity pastures, traditional orchards, hedgerows, wild flower strips, and low intensity cropping strips)
- Crop rotation: at least four different crops have to be cultivated per year on those farms where arable land area exceeds 3 ha and maximum shares of individual crops must be respected
- Soil protection: field parcels that are harvested before 31 August must be sown with main or cover crops by 15 September so that periodical soil erosion is minimised
- Targeted application of pesticides: restrictions on the use and timing of various herbicides and insecticides, consideration of early warning systems and pest forecasts, frequent tests of sprayers
- Animal welfare: farm animals have to be kept according to legal requirements (compliance with the animal protection ordinance).

These environmental cross-compliance criteria aim to address several environmental objectives including reduction of nitrogen and phosphorus runoff and leaching, soil erosion and sediment runoff, conservation and promotion of farmland biodiversity, reduction of pesticide runoff and residues, and improved animal welfare. The percentage of farms complying with the PEP is almost 100%.

Continuation of policy reforms in 2004-13

In contrast with the 1990s, there were no major changes of the system of direct payments and the system was maintained during the period 2004-13 (AP2007 and AP 2011).

The major policy changes were the continuation of the market deregulation started in the 1990s, mainly the abolition of the sugar and dairy intervention systems and a continued gradual (although not dramatic) reduction of border measures. Also export subsidies for primary products were phased out, while export subsidies for some processed products were maintained.

Changes in border and trade measures

One of the main drivers for the changes in the border measures was the agricultural agreement between the European Union and Switzerland which came into force on 1 June 2002. This agreement facilitates market access for both parties. It includes on the one hand, a reduction or complete abolition of import duties on certain products, and on the other hand, simplified commercial procedures.

Milk and dairy products: There have been few changes regarding market access for dairy products. MFN Tariffs on most dairy products remain high, averaging an estimated 101.5%. Switzerland's aggregate tariff-quota commitment for dairy products (527 000 tonne milk equivalent) is divided into six sub-quotas with very small allocations for butter (100 tonnes) and whole milk powder (300 tonnes), which are in surplus. The butter and milk powder quotas are auctioned and no longer subject to *prise en charge* as an allocation method. According to the authorities, the change aims to increase the competition between importers as under the *prise en charge* system only few importers qualify for allocations.

Trade in cheese between Switzerland and the EU has been fully liberalized since June 2007; only a certificate of origin is required. From then on, it was possible for Switzerland and all EU countries to import and export all types of cheese with no restrictions on quantities and no import duty. An evolutionary clause allows for the agreement to be modified in the future. From the Swiss perspective, this is the core of the agricultural agreement, as around 40% of milk production is processed in cheese (milk itself represents around a quarter of total agricultural production). Cheese is also the main exported agro-food commodity.

Meat and livestock: There have been few changes regarding market access for meat and livestock products. Tariffs on most products remain high, averaging an estimated 125.5%. Switzerland's WTO tariff-quota commitments for "red meat" (22 500 tonnes) and "white meat" (54 500 tonnes) are administered through 11 sub-quotas for different categories of meat and meat products. By 2007, auctioning of the sub-quotas had been fully phased in, except for 10% beef and calf sub-quotas.

Feed grains and oilseeds: Switzerland maintains a complex scheme of variable tariffs for feed grains and oilseeds. The basic structure of the scheme has remained unchanged over the years. Tariffs are adjusted periodically so that the duty-inclusive prices are raised to the level of predetermined import prices (*threshold prices* or indicative import values). On the basis of the threshold prices for 11 groups of products, the Federal Department of Economic Affairs determines indicative import values for "similar" products. The statutory threshold prices are reviewed from time to time. In 2007, the threshold price for feed grains (barley) and soyabean meal was reduced by CHF 30/t to CHF 400/t and CHF 470/t respectively, in order to reduce the feed costs to domestic livestock producers (mainly pig and poultry). In 2009, these threshold prices for feed grains (barley) were further reduced to CHF 360/t for feed grains (barley) and CHF 450/t for soyabean meal. The variable tariffs may not exceed the Uruguay Round bindings. There are no tariff quotas for the products covered in the scheme. Protection of the domestic feed industry through tariff escalation was phased-out. As of 1 July 2011, industrial protection elements incorporated in the tariffs for feedstuff mixtures were removed.

From 2008, Switzerland has introduced a new variable tariff regime for cereals of bread-making quality (e.g. wheat). The system is similar to the feed grains scheme (see above). Tariffs are reviewed on a quarterly basis and adjusted, as appropriate (increased or decreased), to stabilize the duty-inclusive price for bread cereals around a minimum import price (*reference price*). The reference price system covers the same tariff lines as the bread-wheat tariff quota. The reference price is reviewed periodically. Tariffs are based on the world market price for cereals, i.e. the c.i.f. price as determined by the Federal Office for Agriculture. The reference price was reduced from CHF 600/tonne to CHF 560/tonne in July 2009. Since 2010, the variable tariffs amount to 100% (up from 60%) of the difference between the world market and the reference price.

Sugar: Switzerland's bilateral agreement with the EU on processed agricultural products (2005) involves bilateral free trade in sugar-containing products. Import duties for sugar used in processed products, and export subsidies between the parties have been phased out (double zero solution). In effect, this agreement involves sugar price levels in Switzerland that are comparable to those of the EU, in order not to compromise the competitiveness of the Swiss food industry. To achieve approximate price parity with the EU, Switzerland has introduced a variable tariff mechanism for MFN sugar imports. MFN tariffs are adjusted, normally every three months, so that duty-inclusive prices are aligned with EU sugar market prices (with a tolerance band of +/- CHF 30/tonne).

Preferential access to Least Developed Countries

Preferential tariff rates are applied to imports from developing countries (LCDs) under a system of preferences schemes. On 1 January 2002, import duty on all agricultural products supplied by these countries was reduced by 30%, and reduced gradually in the following years. Since September 2009, the Swiss government grants zero tariffs on all products imported from least developed countries (LDC), and all agricultural imports are duty and quota free.

Export subsidies

Since 2000, Switzerland has gradually reduced its export subsidies granted to agricultural products. All export subsidies for basic agricultural products were phased-out at the end of 2009. Nevertheless, Switzerland compensates the price handicap of exported processed products due to higher prices of incorporated domestic basic agricultural products (such as milk products, wheat flour or eggs) through a system of import duties and price compensation mechanism for processed agricultural products. Export refunds under this scheme have been phased out for eggs in 2012, but maintained for the rest.

Deregulation of domestic intervention systems

Milk and dairy: In 2003, the Swiss parliament decided to gradually phase out the milk quota system by 2009. In addition, it approved legislation by which production-based milk subsidies may be converted into direct payments. The milk quota and related guaranteed price system was abolished for all dairy farmers as of 1 May 2009, following a transition period from 1 May 2006 to 30 April 2009. Since 1 May 2009, all dairy farmers are obliged to conclude milk delivery contracts with their milk purchasers. The obligation remains in force until 30 April 2015; farmers who sell their milk directly to final consumers and farmers who produce cheeses and other dairy products on farm are exempt.

According to the Law on Agriculture (Article 8), industry and producer organizations are responsible for the marketing of their products. IP LAIT (the Swiss dairy inter-branch organization) has implemented the following measures for its members:

1. A standard contract: on 7 June 2013, the Federal Council approved a request by IP LAIT to declare the standard contract of IP LAIT for the purchase and sale of raw milk and the regulations for the implementation of the market segmentation obligatory for the non-members of IP LAIT (effective from 1 July 2013 until 30 June 2015). All buying and selling of raw milk requires a mandatory contract (minimum duration one year, agreement on price and quantity, segmentation of contractual quantity into A, B or C. Milk traders and processors need to report monthly on the quantity of bought and sold milk per segment (A, B, and C) and on produced and exported dairy products from B and C segments. After a period of 12 months IP LAIT checks how the quantities of bought milk at the B & C segment correspond to the quantity of dairy products made out of B & C milk for selling and exporting and imposes sanctions if significant divergence is found.
2. Milk pricing policy ("recommended" prices) based on market segmentation: i) the "A segment" comprises domestic sales of milk products (CHF 0.68/kg of milk in 2013); ii) the "B segment" includes world market exports of skimmed-milk powder (milk protein) and domestic sales of corresponding butter production (milk fat) (CHF 0.61/kg in 2013); iii) the "unsupported C segment" comprises world market exports of butter and SMP (CHF 0.40/kg in 2013). The milk buyers have committed to buy at least 60% of their raw milk in the A-segment. In 2013 (annual average) the milk delivered from Swiss producers was distributed as follows: A segment 89%; B segment 10.7% and C segment 0.3%.
3. Compulsory levy: on 31 August 2011, the Federal Council approved a request by IP LAIT to implement a compulsory levy of CHF 0.01/kg on milk deliveries by non-members of IP LAIT for market relief measures in the milk-fat market (effective until 30 April 2013). The rationale was to prevent free-riders from undermining the IP LAIT initiative to stabilize the market; non-members of IP LAIT represent about 5% of processed milk in Switzerland. Funds from members and non-members of IP LAIT, about CHF 65 million, were used to support sales of surplus butter and SMP to the world market (unsupported C segment).

Since 2010 price support payments for dairy products consist only of the payments per tonne for milk transformed into cheese and an additional output payment if milk is produced without silage feed. Due to border measures the price paid to milk producers remains on average above (but getting closer to) the world market prices.

Sugar: Under the Sugar Act (*Ordonance sur le sucre* 916.114.11) from 1998, a sugar regulations regime was applied. Under this regime the Federal authorities (OFAG) makes an agreement with *SA Sucreries Aaberg et Frauenfeld – SAF* (a society regrouping the only two sugar refineries in Switzerland) to produce annually a minimum of 150 000 tonnes of sugar from domestic sugarbeet, and to negotiate with the Federation of sugarbeet growers the prices and the quantities delivered by individual producers at those prices. The SAF receives an annual subsidy for processing domestic sugarbeet. The c.i.f. price of imported sugar is taken into account when calculating the subsidy for each season. By 1st October 2009 the Sugar Act was abolished as well as the subsidy paid to processors. However, the sugarbeet quotas and prices continue to be agreed on a private basis between the SAF and the Federation of sugarbeet growers. As compensation for lower prices for sugarbeet due to the sugar market reform in the EU (2006-09), an area payment for sugarbeet was introduced in 2008.

Changes in the direct payments

As indicated earlier, there were no significant changes in the system of direct payments. Apart from the introduction of area payments for sugarbeet (see above), a new programme was introduced in the Ecological direct payments in 2008. A programme *Sustainable use of agricultural resources* provides financing for 6 year projects developed by local authorities designed to improve the use of natural resources in specific areas. These programmes are co-financed by the Federal budget (maximum 80% of costs) and by the Cantons (minimum 20% of costs).

Forthcoming policy AP 2014-17

Switzerland has adopted a new policy framework for 2014-17 (*Politique Agricole* 2014-2017). The main policy objectives are the same as in the previous period i.e.: food security (maintain self-sufficiency at the current level of around 60%); efficient and sustainable use of natural resources; cultural landscape; and support to less favoured areas.

The policy reform focuses on a re-arrangement and fine-tuning of the *direct payment* scheme, intended to improve the efficiency and effectiveness of the measures, and set up a system of direct payments linked to the various objectives. The main change is the suppression of general area payments and reallocation of payments more closely related to specific objectives (agricultural practices) complemented by a system of transition payments to make the reform socially acceptable. Another important shift is the replacement of general headage payments to ruminants by an area payment to pastures with a requirement for a minimal stocking density. The environmental cross-compliance conditions are maintained in the new system of payments. The overall budgeted annual amount of these payments remains stable for the whole period around CHF 2.8 billion, which is around the same level as in 2012 and 2013.

The revised *direct payments* scheme has seven categories, which are linked to the achievement of specific policy objectives and the provision of public goods:

1. **Payments for ensuring food supplies** (i.e. food-security payments): consist mainly of area payments, with rates differentiated between the plain and hilly and mountain regions. The payments for production in difficult conditions are also part of this category.
2. **Farmland payments**: are also area payments with a main function to maintain extensive forms of agricultural production in especially difficult conditions to maintain a cultivated landscape;
3. **Bio-diversity payments**: are more output oriented payments targeted to specific outcomes or farming practices; Especially the quality of the ecological compensation areas will be enhanced to improve the habitat and the possibilities for dispersal of target and indicator species in agriculture;
4. **Payments for landscape quality**: payments for preservation and promotion of landscape diversity (including more diverse crop rotation, flowering fields, and traditional agricultural practices), based on local projects and co-financed by the Cantons;

5. **Payments for production systems**: area and headage payments to provide incentives for environment/animal-friendly production systems (e.g. organic farming).
6. **Resource-efficiency payments**: payments providing incentives to use specific production techniques (e.g. certain manure spreading methods and soil conservation methods like no-till).
7. **Transitional payments**: are provided to farmers who suffer a loss of direct payments under the new system. These payments are scheduled to decrease gradually.

The system is complex and each category includes several programmes. These programmes are a combination of new programmes and "old" programmes, i.e. already implemented under the AP 2011 package. Box 3.2 provides more detailed information on the programmes in the main categories of the AP 2014-17.

Box 3.2. System of direct payments introduced under the AP 2014-17

1. Payments for ensuring food supplies

- *Basic Contribution* (new): is a general area payment which replaces the headage payment to ruminants. This shift sets at the same level the payment for arable crops and grassland (the previous system privileged the grassland areas).
- *Contribution to production in difficult conditions* (new): an area payment provided to farms in difficult conditions and replaces the headage payments for animals raised in difficult conditions (by definition this payment is for the mountain and hilly areas).
- *Contribution to arable land and perennial cultures* (new): an additional area contribution to crops on arable land and perennial cultures.

2. Farmland payments

- *Contribution to maintain an open landscape (new)*
- *Farming on steep slopes (old)*: area payments for farming in specifically defined conditions
- *Farming on very steep slopes* (new): area payments for farming in specifically defined conditions
- *Wine production on steep slopes* (old): area payments
- *Alpine pasturing* (new)
- *Summer pasturing* (old)

3. Bio-diversity payments

- *Contribution to environmental quality level 1* (old): regroups the payments provided under the various programmes provided under Ecological compensation in the previous system.
- *Contribution to environmental quality level 2* (old): corresponds to the payments provided under the Ecological quality directive in the previous system.
- *Contribution to environmental quality level 3* (new): these payments will be provided from 2016 to finance projects listed as objects of national importance.
- *Payments for ecological compensation areas* (old)
- *Payments for creating networks of highly valuable biodiversity areas* (new)

4. Payments for landscape quality

- Contribution for quality of typical regional landscapes (new): These projects are developed by Cantons and are co-financed from Federal and Cantonal budgets.

Box 3.2. **System of direct payments introduced under the AP 2014-17** *(cont.)*

5. Payments for production systems

- *Payments for organic farming (old)*
- *Payments for extensive production (grains and rapeseed) (old)*
- *Payments for animal welfare:* i) *payments for regularly keeping animals outdoors* (old) and ii) *animal welfare through housing systems* (old)
- *Contribution to meat and milk production on grassland* (new): grassland based area payments conditional to minimal stocking densities and restricted use of concentrated feed.

6. Resource-efficiency payments

- *Contribution to spreading techniques limiting the emission of pollutants (new)*
- *Contribution to cultivation techniques preserving the soils (new)*
- *Contribution to precision application of phytosanitary products (new)*
- *Contribution to water protection (art.62) (old)*
- *Contribution to sustainable use of resources (art.77a/b) (old)*

7. Transitional payments (new)

Support to agriculture

The OECD estimates the level of support to agriculture for all OECD countries and an increasing range of emerging economies which are key players on the world markets. The OECD provides estimates of support to farming for countries which represent 80% of world value added in agriculture. For these countries the OECD applies the methodology of Producer Support Estimate (PSE) which provides consistent and comparable information on the level and structure of support to agriculture.

Estimates of support, since 1986, are also provided for Switzerland and the Swiss data are included in the OECD PSE/CSE Database. This section provides an assessment of how the reforms implemented since the early 1990s have transformed the level and structure of support, based on the various OECD indicators of agricultural support.

Overall support and its structure

The aggregate support

Although the level of support in Switzerland, as measured by the PSE, has declined gradually following the implementation of reforms started in the 1990s, it is still one of the highest amongst OECD countries. In the mid-1990s around 70% of Swiss gross farm receipts were from public transfers either from consumers or taxpayers, this share is around 50% in 2011-13. However, Switzerland is still a country with one of the highest level of support as measured by the percentage PSE together with Japan, Korea and its fellow EFTA countries Norway and Iceland (Figure 3.4). The 1995-97 average ranks Switzerland first, and the 2011-13 average third among OECD countries.

This level of support measured in 2011-13 is also more than two times higher than in the European Union (EU), which is the closest regional area and also the main trading partner in agro-food products. It is also two times higher than the OECD average (Figure 3.4).

Figure 3.4. **Producer Support Estimate by country, 1995-97 and 2011-13**

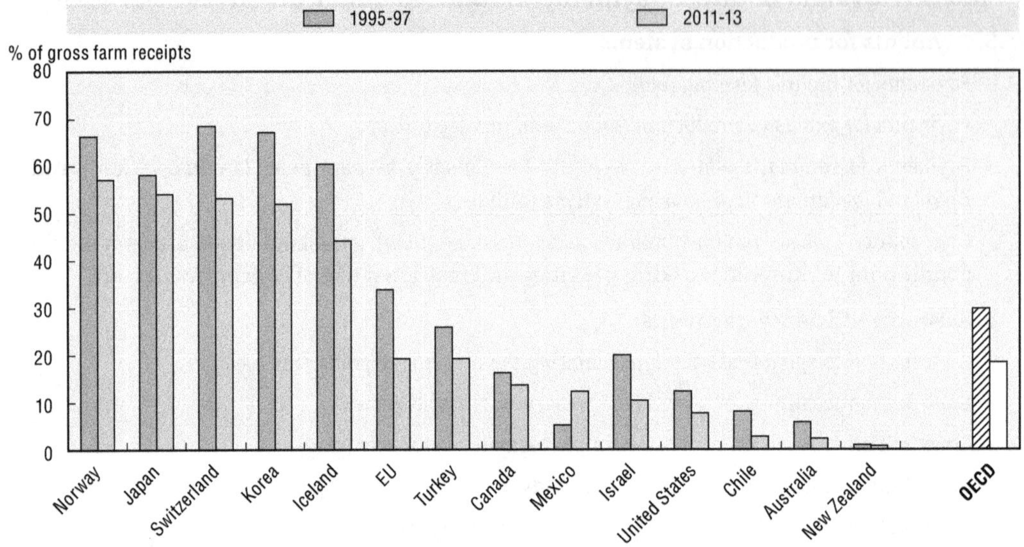

Source: OECD (2014), "Producer and Consumer Support Estimates", *OECD Agriculture statistics* (database).

In the period 1986-2013, the level of support shows a downward trend from three-quarters of gross farm receipts in 1986 to around a half in 2013. The yearly fluctuations in the level of support are mainly driven by the developments of the support based on commodity output which is mainly the Market Price Support which reflects also the fluctuations in the world prices and the exchange rate (Figure 3.5).

Figure 3.5. **Switzerland: Producer Support Estimate level and composition by support categories, 1986-2013**

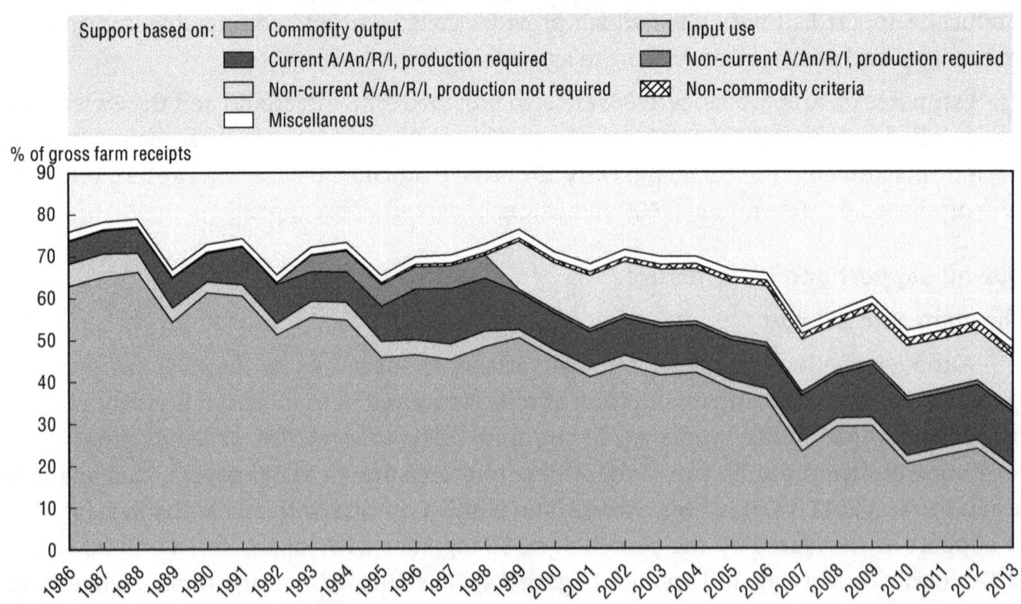

Source: OECD (2014), "Producer and Consumer Support Estimates", *OECD Agriculture statistics* (database).

Mainly the changes in the structure of support (more than its level) illustrate the implementation of the policy reforms since the early 1990s, described in the previous section of this Chapter.

Support by main PSE categories

Support based on commodity output (category at the bottom of Figure 3.5): this category represents the major shift in the PSE, which is mainly due to the continuous reduction of the Market Price Support (MPS) which is the main element of the support based on commodity output. This development clearly illustrates the deregulation of domestic market and reduction of the tariff barriers, which resulted into a decline of domestic prices. The annual fluctuations in the MPS are mostly driven by fluctuations in world market prices and exchange rates and reflect the isolation of the domestic market prices from these effects. The only important element of payments based on output is the payments for milk for cheese production.

Support based on input use: this support, based on variable input use and fixed capital formation, represents a relatively small and declining share of support. The main element of the support based on variable input use was the *Feed grain price reduction*, which has been gradually reduced since 1994 and discontinued in 2000. The other major element in this category is the investment credit subsidy provided to farmers across the whole period.

Payments based on current area and animal numbers: this type of payment represents consistently a rather important part of the overall support and at the end of the analysed period it is about one-third of the total support. This category combines payments provided under various programmes applied under different policy reforms. In the earlier period these were acreage premiums for wheat, coarse grains, oilseeds and potatoes and headage payments for cows without milk production and for livestock in mountain areas. The role of the headage payments increased during the 1990s with the introduction of general payments for ruminants and payments provided under animal welfare programmes. In the period from 1993 to 1998, the payments for integrated production were an important part of this category.

Payments based on historical (non-current) parameters (Area/Animal numbers/Revenue/Income) production required: this category consists mainly of the Complementary Direct Payments applied during 1993-1998. More specifically this category includes two elements of these payments – the *Base farm payments* and the *Supplementary payments*. The other two elements Base Area Payment – arable land and Base Area Payment – grassland are in the category *Payments based on current area and animal numbers*.

Payments based on historical (non-current) parameters (Area/Animal numbers/Revenue/Income) production not required: this category includes only the General Area payment introduced in 1999 and which replaced the Complementary Direct payments (see above). However, this programme carries important amount of payments and represented around 30% of total support in 2013.

Payment based on non-commodity criteria: this category, although increasing, remains relatively marginal in the overall support. It includes payments provided under some programmes from the Ecological payments such as: Extensive meadows, Floral fallow, Extensive area strips, Hedges and rustic groves and Tall fruit trees. It includes also payments for Contribution to environmental quality.

Characteristics of policy support

Commodity specificity

The labels included in the PSE database also provide the information on commodity specificity of the individual support programmes, i.e. which payments are commodity specific and to which commodity they belong. The analysis in this part illustrates mainly the effect of market deregulation (terminated in 1999) and the reduction of tariff barriers in the latter period. As the policies implemented were moving away from market price support and commodity specific payments to more general payments, the share of payments to specific commodities – the Single Commodity Transfer (SCT) – has declined from 86% of total support in 1986-88 to 69% in 1995-97 and to 40% in 2011-13 (Figure 3.6).

Figure 3.6. **Single Commodity Transfer by commodity**

Percentage of commodity gross farm receipts

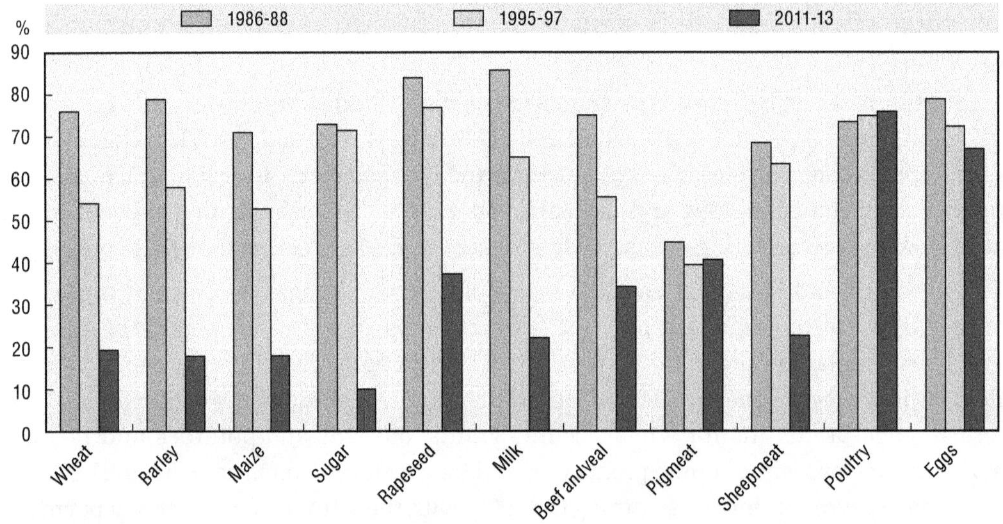

Source: OECD (2014), "Producer and Consumer Support Estimates", *OECD Agriculture statistics* (database).

The main indicators providing information for specific commodities are:

1. *The percentage SCT* – indicates the share of single commodity transfers on commodity gross farm receipts for the specific commodity.
2. *Producer Nominal Protection Coefficient* (producer NPC): the ratio between the average price received by producers (at farm gate), including payments per tonne of current output, and the border price (measured at farm gate).
3. *Producer Nominal Assistance Coefficient* (producer NAC): the ratio between the value of gross farm receipts including support and gross farm receipts (at farm gate) valued at border prices (measured at farm gate).
4. *Consumer Nominal Protection Coefficient* (consumer NPC): the ratio between the average price paid by consumers (at farm gate) and the border price (measured at farm gate). The Consumer NPC is also available by commodity.
5. *Consumer Nominal Assistance Coefficient* (consumer NAC): the ratio between the value of consumption expenditure on agricultural commodities (at farm gate) and that valued at border prices.

The analysis of these indicators shows a large shift from commodity specific support, mainly due to a substantial reduction of the domestic prices and resulting reduction of the price gap between domestic and world market prices. In the 1980s the domestic prices were on average 4.5 times higher than world prices and SCTs represented around 70-80% of gross farm receipts for the specific commodity. In the mid-1990s the price gap and the SCT share in total receipts was reduced with the exception of poultry and eggs. This trend continues in the following period and in most recent years (2011-13) the prices went much closer to world market levels. The average NPC at 1.4 indicates that prices are on average 40% above the world prices (Figure 3.7), but there is a huge variation among individual commodities. A detailed overview of the percentage SCT and the Producer NPC by commodity is provided in Figures 3.6 and 3.8; Annex Table 3.A1.2.

Figure 3.7. **Producer Nominal Protection Coefficient (PNPC) and Producer Nominal Assistance Coefficient (PNAC), 1986-2013**

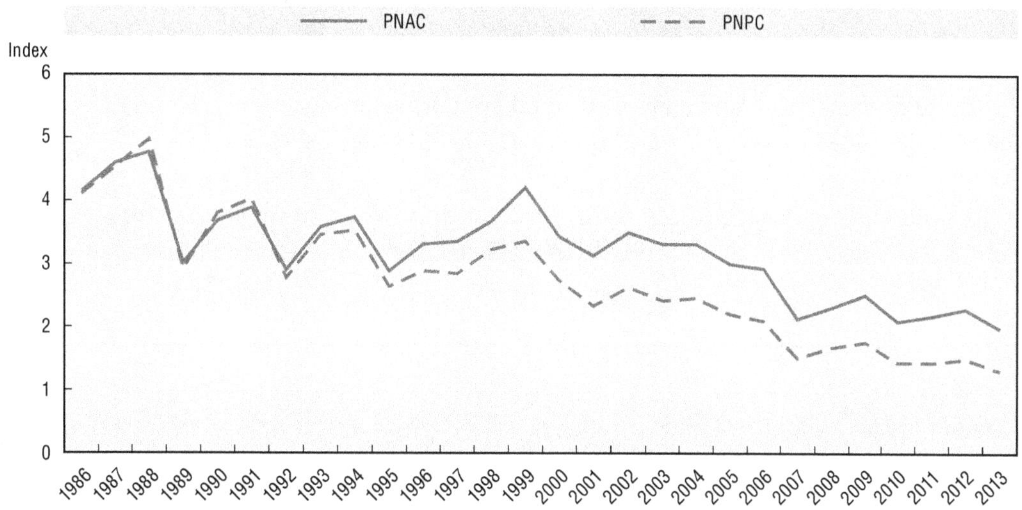

Source: OECD (2014), "Producer and Consumer Support Estimates", *OECD Agriculture statistics* (database).

Figure 3.8. **Producer Nominal Protection Coefficient by commodity**

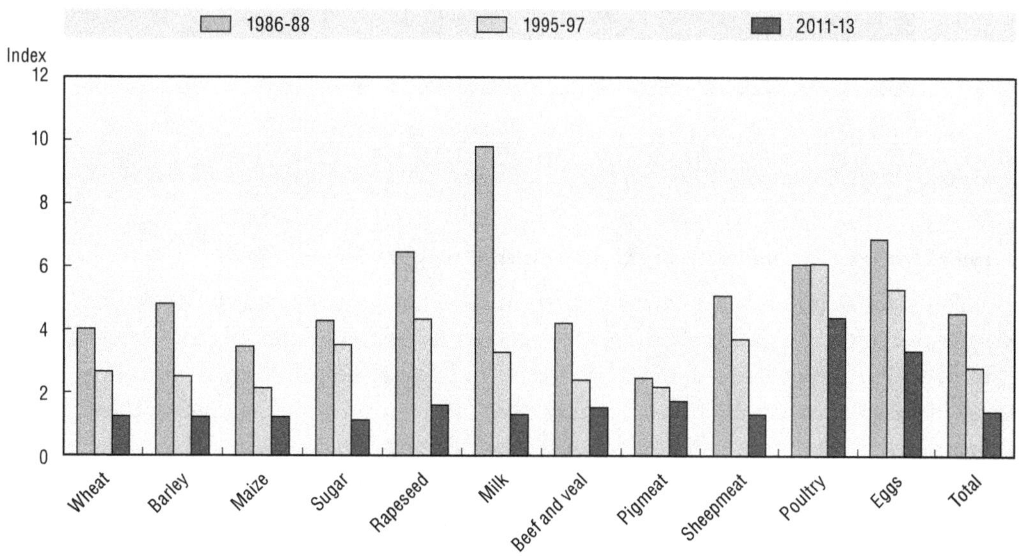

Source: OECD (2014), "Producer and Consumer Support Estimates", *OECD Agriculture statistics* (database).

Similarly the development of the Consumer NPC indicates a reduced price gap to consumers measured at the farm gate level, which indicates a reduced implicit taxation of consumers. A detailed overview of the Consumer NPC by commodity is provided in Annex Table 3.A1.3.

Production and trade distorting forms of support

Another important angle of analysis is to what extent the policies applied are production and trade distorting. In the PSE classification the payments based on output (including the MPS) and the payments based on variable input use (without input constraints) are identified as those with a closest link to producers decisions and hence potentially most production and trade distorting.

In the Swiss policy mix these policies are represented mainly by the market price support and by payments based on output for milk for cheese processing. Due to a significant reduction of the MPS from the high levels in the 1980s, the share of the most production and trade distorting policies has been gradually reduced (Figure 3.9). In 1986-88 the share of most distorting support in total support to producers represented 89%, this share was reduced to 69% in 1995-97 and to 41% in 2011-13.

Figure 3.9. **Share of the most production and trade distorting form of support in the PSE**

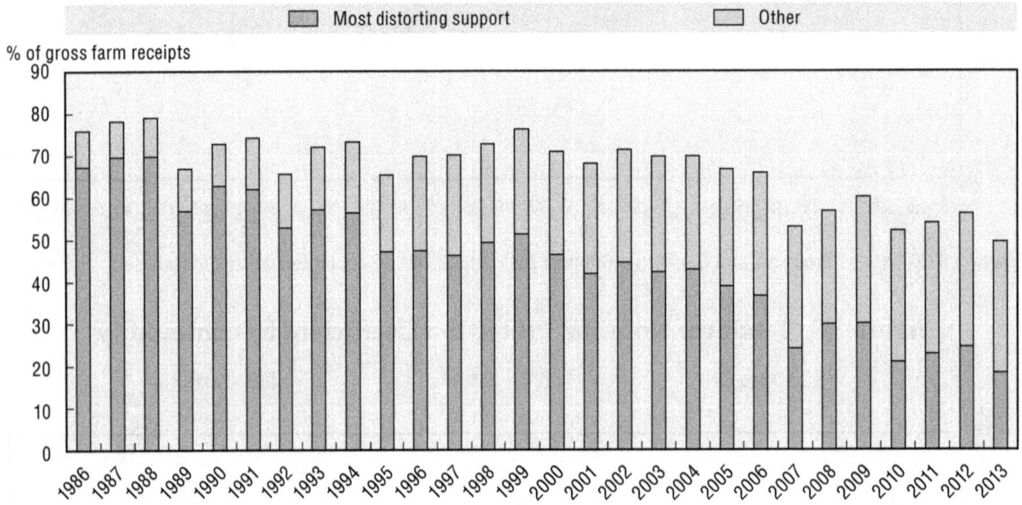

Source: OECD (2014), "Producer and Consumer Support Estimates", *OECD Agriculture statistics* (database).

Support targeted at environmental and animal welfare issues

The agricultural policy reforms started in 1993 have introduced the Ecological Direct Payments, which is a group of payments received by farmers to apply environmental and animal welfare standards above those required by regulations on a voluntary basis. Although those payments represent a minor part of the total direct payments, they have an upward trend and their share in total payments and total support is increasing (Figure 3.10). It is worth noting, that the animal welfare payments are almost as high that the agri-environmental payments. As the animal welfare payments are paid per head of livestock, their incidence on production decisions is higher than for most of the agri-environmental payments.

The General Services Support Estimate and Total Support Estimate

In contrast with the PSE, which is an indicator of transfers going directly to individual farms, the General Services Support Estimate (GSSE) captures public expenditures that contribute to create enabling conditions for the primary agricultural sector through

Figure 3.10. **Share of the environmental and animal welfare payments in total payments**

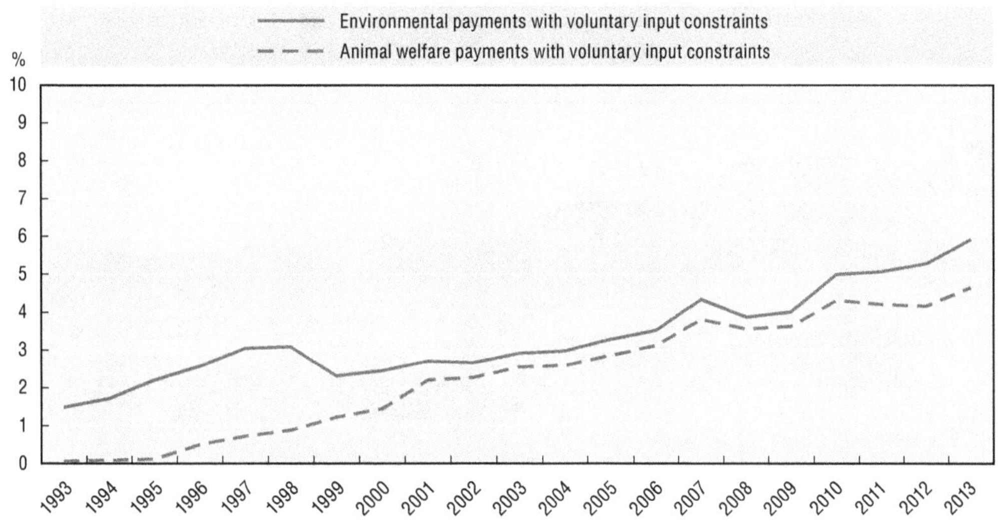

Source: OECD (2014), "Producer and Consumer Support Estimates", *OECD Agriculture statistics* (database).

development of private or public services, institutions and infrastructure. The GSSE transfers do not directly alter producer receipts or costs or consumption expenditures, although they may over time indirectly affect production or consumption of agricultural commodities. The GSSE expenditure typically includes financing of activities such as the Agricultural knowledge and information system, Inspection and control, development and maintenance of infrastructure, marketing and promotions, etc.

In nominal terms, the GSSE expenditure recorded a downward trend in the period 1986-2013. It went down from a yearly expenditure of CHF 677 million in 1986-88, to CHF 590 million in 1995-97 and to CHF 500 million in 2011-13. However, in relative terms, as a share of the Total Support Estimate (TSE), the percentage GSSE records an upward trend, especially from the mid-nineties. This is due to a reduction in the PSE, which went at a higher pace than the GSSE reduction. In 1986-88 and 1995-97 the share of GSSE in the TSE was stable at around 6.5%. In the following period it increased to 8.6% in 2011-13. (Figure 3.11, Annex Table 3.A1.1).

In terms of the main categories of the GSSE a reduction of public expenditures was recorded in all categories except marketing and promotion. The sharpest decline was in the category Costs of public stockholding, which is related to the discontinuation of the heavy intervention system during the 1990s.

The Total Support Estimate (TSE) regroups the transfers in the PSE, the GSSE spending and the eventual budgetary payments to consumers. In nominal terms, the TSE recorded a downward trend in the period 1986-2013. It decreased from a yearly expenditure of CHF 10.3 billion in 1986-88, to CHF 9 billion in 1995-97 and to CHF 5.8 billion in 2011-13. In relative terms the percentage TSE is expressed as a share of GDP, to indicate the relative burden to the economy. In these relative terms the decline of the TSE was more pronounced, as it decreased from 3.7% in 1986-88 to 2.3% in 1995-97 and 1% in 2011-13 (Annex Table 3.A1.1). However, the main driver in this relative indicator was not the reduction of the TSE itself, but the dynamic development of the other part of the economy, especially the service sector.

Figure 3.11. **Level and structure of the General Services Support Estimate**

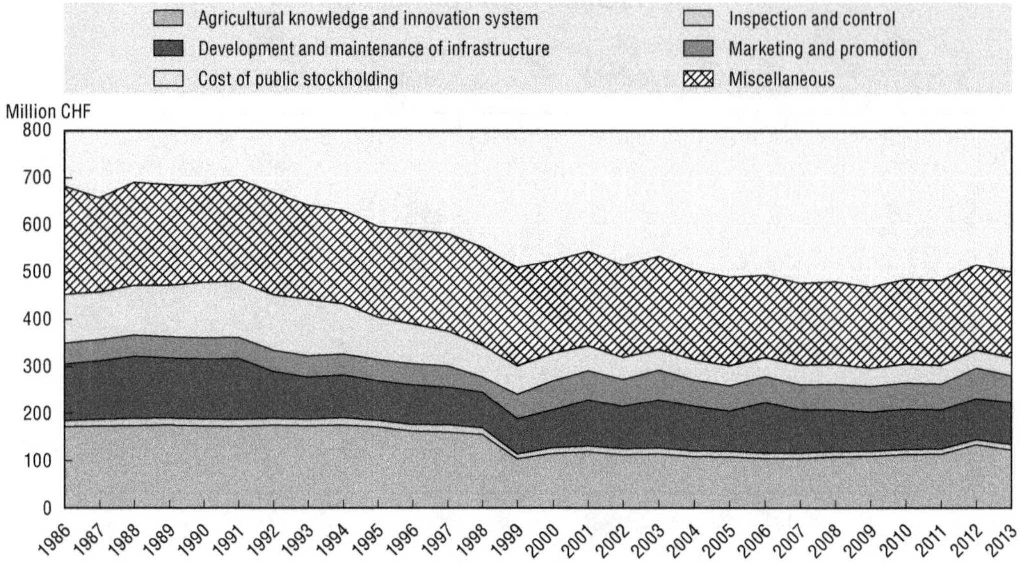

Source: OECD (2014), "Producer and Consumer Support Estimates", *OECD Agriculture statistics* (database).

ANNEX 3.A1

Detailed indicators of support to agriculture

Table 3.A1.1. **Switzerland: Estimate of support to agriculture**

Million CHF

	1986-88	1995-97	2011-13	2011	2012	2013p
Total value of production (at farm gate)	9 482	8 236	6 521	6 586	6 404	6 574
of which: share of MPS commodities (%)	81.5	82.3	69.9	71.2	71.4	67.1
Total value of consumption (at farm gate)	11 394	9 557	7 899	7 902	7 746	8 048
Producer Support Estimate (PSE)	8 509	7 362	5 330	5 442	5 566	4 983
Support based on commodity output	7 091	4 918	2 118	2 230	2 356	1 769
Market Price Support[1]	7 049	4 835	1 822	1 938	2 058	1 470
Payments based on output	42	83	296	292	298	299
Payments based on input use	563	411	201	198	201	203
Based on input use	454	309	81	81	81	81
with input constraints	0	180	14	14	13	14
Based on fixed capital formation	72	78	119	116	119	121
with input constraints	0	0	0	0	0	0
Based on on-farm services	36	25	1	1	1	1
with input constraints	0	0	0	0	0	0
Payments based on current A/An/R/I, production required	612	1 203	1 310	1 309	1 310	1 311
Based on Receipts/Income	15	0			0	0
Based on Area planted/Animal numbers	597	1 203	1 310	1 309	1 310	1 311
with input constraints	340	1 050	1 299	1 297	1 299	1 299
Payments based on non-current A/An/R/I, production required	28	569	102	102	102	102
Payments based on non-current A/An/R/I, production not required	0	0	1 203	1 218	1 195	1 195
With variable payment rates	0	0	0	0	0	0
with commodity exceptions	0	0	0	0	0	0
With fixed payment rates	0	0	1 203	1 218	1 195	1 195
with commodity exceptions	0	0	0	0	0	0
Payments based on non-commodity criteria	0	61	200	190	205	206
Based on long-term resource retirement	0	0	0	0	0	0
Based on a specific non-commodity output	0	61	200	190	205	206
Based on other non-commodity criteria	0	0	0	0	0	0
Miscellaneous payments	216	200	197	196	198	198
Percentage PSE (%)	77.7	68.4	53.2	53.9	56.2	49.4
Producer NPC (coeff.)	4.54	2.79	1.41	1.44	1.49	1.30
Producer NAC (coeff.)	4.51	3.18	2.14	2.17	2.28	1.98
General Services Support Estimate (GSSE)[2]	677	590	499	483	515	500
Agricultural knowledge and innovation system	173	164	123	114	133	123
Inspection and control	14	15	11	11	11	11
Development and maintenance of infrastructure	126	83	87	83	87	90
Marketing and promotion	45	45	59	55	65	57
Cost of public stockholding	103	83	39	40	38	39
Miscellaneous	216	200	180	180	180	180
Percentage GSSE (% of TSE)	6.6	6.5	8.6	8.1	8.5	9.1
Consumer Support Estimate (CSE)	-7 535	-4 994	-2 133	-2 260	-2 345	-1 792
Transfers to producers from consumers	-7 088	-5 053	-1 660	-1 784	-1 895	-1 301
Other transfers from consumers	-1 767	-1 221	-497	-499	-478	-514
Transfers to consumers from taxpayers	1 099	1 053	7	5	4	11
Other transfers from consumers	-1 767	-1 221	-497	-499	-478	-514
Transfers to consumers from taxpayers	1 099	1 053	7	5	4	11
Excess feed cost	221	227	17	18	23	11
Percentage CSE (%)	-73.1	-58.7	-27.1	-28.6	-30.3	-22.3
Consumer NPC (coeff.)	4.50	2.91	1.38	1.41	1.44	1.29
Consumer NAC (coeff.)	3.74	2.42	1.37	1.40	1.43	1.29
Total Support Estimate (TSE)	10 285	9 005	5 836	5 930	6 085	5 494
Transfers from consumers	8 855	6 274	2 157	2 283	2 373	1 815
Transfers from taxpayers	3 197	3 952	4 177	4 146	4 190	4 194
Budget revenues	-1 767	-1 221	-497	-499	-478	-514
Percentage TSE (% of GDP)	3.7	2.3	1.0	1.0	1.0	0.9
GDP deflator (1986-88 = 100)	100	125	143	142	143	143

Notes:

p: provisional.

NPC: Nominal Protection Coefficient.

NAC: Nominal Assistance Coefficient.

A/An/R/I: Area planted/Animal numbers/Receipts/Income.

1. Market Price Support (MPS) is net of producer levies and excess feed cost. MPS commodities for Switzerland are: wheat, maize, barley, rapeseed, sugar, milk, beef and veal, sheep meat, pigmeat, poultry and eggs.

Source: OECD (2014), "Producer and Consumer Support Estimates", *OECD Agriculture statistics* (database).

Table 3.A1.2. **Switzerland: Producer Single Commodity Transfers**

Million CHF

	1986-88	1995-97	2011-13	2011	2012	2013p
TOTAL						
PSE (million CHF)	8 509	7 362	5 330	5 442	5 566	4 983
Producer SCT (million CHF)	7 294	5 073	2 130	2 241	2 367	1 781
Share of Producer SCT in PSE (%)	85.7	69.0	39.8	41.2	42.5	35.7
Wheat						
Producer SCT (million CHF)	417	333	48	46	58	39
Percentage SCT (%)	76.0	54.1	19.3	17.2	23.1	17.5
Producer NPC (coeff.)	4.02	2.66	1.24	1.21	1.30	1.21
Maize						
Producer SCT (million CHF)	102	63	9	8	13	7
Percentage SCT (%)	70.9	52.8	17.9	15.2	23.7	14.8
Producer NPC (coeff.)	3.46	2.13	1.22	1.18	1.31	1.17
Barley						
Producer SCT (million CHF)	153	102	11	10	14	9
Percentage SCT (%)	78.9	57.9	17.8	15.0	22.7	15.6
Producer NPC (coeff.)	4.80	2.50	1.22	1.18	1.29	1.19
Rapeseed						
Producer SCT (million CHF)	80	57	23	25	21	24
Percentage SCT (%)	83.9	76.8	37.3	36.7	34.6	40.7
Producer NPC (coeff.)	6.45	4.32	1.60	1.58	1.53	1.69
Sugar						
Producer SCT (million CHF)	95	111	12	8	7	20
Percentage SCT (%)	72.9	71.4	9.9	5.4	5.3	19.1
Producer NPC (coeff.)	4.28	3.51	1.12	1.06	1.06	1.24
Milk						
Producer SCT (million CHF)	2 775	2 132	468	479	634	292
Percentage SCT (%)	85.7	65.0	22.1	22.0	30.2	14.1
Producer NPC (coeff.)	9.80	3.27	1.30	1.29	1.45	1.17
Beef and Veal						
Producer SCT (million CHF)	1 312	646	398	449	405	340
Percentage SCT (%)	75.0	55.5	34.3	38.4	35.0	29.3
Producer NPC (coeff.)	4.21	2.40	1.53	1.63	1.54	1.42
Sheepmeat						
Producer SCT (million CHF)	36	42	9	5	9	12
Percentage SCT (%)	68.5	63.4	22.6	13.1	23.9	30.8
Producer NPC (coeff.)	5.08	3.70	1.31	1.16	1.32	1.45
Pigmeat						
Producer SCT (million CHF)	717	458	352	391	363	301
Percentage SCT (%)	44.8	39.4	40.7	44.3	41.1	36.6
Producer NPC (coeff.)	2.45	2.17	1.73	1.83	1.75	1.60
Poultry						
Producer SCT (million CHF)	112	133	122	118	124	124
Percentage SCT (%)	73.5	74.9	75.9	76.6	75.9	75.1
Producer NPC (coeff.)	6.08	6.10	4.38	4.55	4.45	4.15
Eggs						
Producer SCT (million CHF)	185	135	123	134	117	117
Percentage SCT (%)	78.9	72.4	67.1	73.6	63.6	63.9
Producer NPC (coeff.)	6.87	5.28	3.32	4.10	2.97	2.88
Other Commodities[1]						
Producer SCT (million CHF)	1 310	862	555	569	601	495
Percentage SCT (%)	82.0	65.9	32.0	33.6	37.1	25.3
Producer NPC (coeff.)	4.50	2.90	1.27	1.30	1.34	1.18

Notes:

p: provisional.

SCT: Single Commodity Transfers.

PSE: Producer Support Estimate.

NPC: Nominal Protection Coefficient.

1. Producer SCT for Other Commodities: Total Producer SCT minus the Producer SCTs for the commodities listed above.

Source: OECD (2014), "Producer and Consumer Support Estimates", *OECD Agriculture statistics* (database).

Table 3.A1.3. **Switzerland: Consumer Single Commodity Transfers**

Million CHF

	1986-88	1995-97	2011-13	2011	2012	2013p
TOTAL						
CSE (million CHF)	-7 535	-4 994	-2 133	-2 260	-2 345	-1 792
Consumer SCT (million CHF)	-7 749	-5 115	-2 136	-2 263	-2 348	-1 796
Wheat						
Consumer SCT (million CHF)	-538	-399	-87	-75	-104	-82
Consumer NPC (coeff.)	4.02	3.10	1.24	1.21	1.30	1.21
Maize						
Consumer SCT (million CHF)	-139	-32	-12	-10	-16	-11
Consumer NPC (coeff.)	3.46	2.13	1.22	1.18	1.31	1.17
Barley						
Consumer SCT (million CHF)	-207	-44	-11	-9	-13	-12
Consumer NPC (coeff.)	4.80	2.50	1.22	1.18	1.29	1.19
Rapeseed						
Consumer SCT (million CHF)	-313	-252	-132	-134	-123	-139
Consumer NPC (coeff.)	6.45	4.32	1.60	1.58	1.53	1.69
Sugar						
Consumer SCT (million CHF)	-143	-146	-17	-9	-10	-33
Consumer NPC (coeff.)	4.51	3.51	1.12	1.06	1.06	1.24
Milk						
Consumer SCT (million CHF)	-1 900	-1 102	-175	-189	-336	0
Consumer NPC (coeff.)	9.85	3.27	1.12	1.12	1.24	1.00
Beef and Beal						
Consumer SCT (million CHF)	-1 382	-712	-442	-502	-448	-375
Consumer NPC (coeff.)	4.21	2.40	1.53	1.63	1.54	1.42
Sheepmeat						
Consumer SCT (million CHF)	-106	-102	-20	-11	-21	-27
Consumer NPC (coeff.)	5.08	3.70	1.31	1.16	1.32	1.45
Pigmeat						
Consumer SCT (million CHF)	-908	-651	-373	-417	-386	-317
Consumer NPC (coeff.)	2.45	2.17	1.73	1.83	1.75	1.60
Poultry						
Consumer SCT (million CHF)	-301	-298	-237	-236	-239	-234
Consumer NPC (coeff.)	6.08	6.10	4.38	4.55	4.45	4.15
Eggs						
Consumer SCT (million CHF)	-399	-299	-235	-261	-224	-220
Consumer NPC (coeff.)	6.87	5.28	3.32	4.10	2.97	2.88
Other Commodities[1]						
Consumer SCT (million CHF)	-1 414	-1 079	-395	-409	-427	-349
Consumer NPC (coeff.)	4.34	2.99	1.25	1.27	1.30	1.18

Notes:
p: provisional.
SCT: Single Commodity Transfers.
CSE: Consumer Support Estimate.
NPC: Nominal Protection Coefficient.

1. Consumer SCT for Other Commodities: Total Consumer SCT minus the Consumer SCTs for the commodities listed above.

Source: OECD (2014), "Producer and Consumer Support Estimates", *OECD Agriculture statistics* (database).

OECD Review of Agricultural Policies: Switzerland 2015

Chapter 4

Impact of Swiss policy reforms on the economic and environmental performance of agriculture

This chapter assesses the impact of Swiss agricultural policy reforms on the economic and environmental performance of agriculture. OECD Policy Evaluation Model (PEM) is used to explore the impacts of policy reforms on production, trade and farm income. This Chapter also attempts to assess the environmental impacts of policy reforms, such as impacts on nutrient balances and surpluses, greenhouse gas emissions and biodiversity, making use of OECD environmental indicators and other data sources. Given the importance of sustainability and landscape objectives of the agricultural policies, the study also explores the use of PEM to assess the environmental impacts of agricultural and agri-environmental policies by incorporating spatial heterogeneity.

Assessment of past agricultural policy reforms in Switzerland

The gradual and continuous reforms since the early 1990s have reduced market price support while providing complementary direct payments to farmers. Simultaneously, policies have addressed the environmental performance of agriculture through regulation, cross-compliance conditions attached to direct payments and specific ecological payments. To assess the economic and environmental impacts of the reduction and re-orientation of farm support, this chapter disentangles the effects of policy changes from exogenous factors that have affected the sector over the past two decades. The analysis highlights the effects of policy incentives that have led to notable changes in the volume and structure of production, emphasising livestock production in mountainous region while reducing the profitability of crops production in plain regions, and discusses the concomitant environmental effects.

To isolate the effects of policies, the analysis employs the OECD's Policy Evaluation Model (OECD PEM), see Box 4.1. The Swiss module of the OECD PEM has been augmented in two ways: since the production characteristics as well as the policy design are intimately linked to geographical conditions of production, the Swiss module distinguishes three regions: plain, hilly and mountain regions.[1] Second, the augmented module contains an accounting of key environmental performance indicators.

Box 4.1. **The OECD Policy Evaluation Model**

The Policy Evaluation Model (PEM) was developed by the OECD to relate the information in the PSE database to specific economic outcomes. It is designed to complement the information in the PSE database with a basic economic structure and underlying data to allow aggregate representations of seven OECD countries (Canada, Mexico, Japan, Korea United States and Switzerland, and the European Union as a single region with two production zones, the EU15 and EU12).

The PEM provides a static, partial equilibrium view of the impact of agricultural policies, with elasticities intended to represent a medium-term adjustment of approximately five years. The PEM models aggregate production of wheat, coarse grains, oilseeds, rice, milk, and beef that are produced with a combination of farm-owned factors of production and purchased inputs.

The PEM is an equilibrium displacement model; that is, it is calibrated such that for each year all markets are in equilibrium given observed data on prices, quantities, and support. Policy simulations disturb this equilibrium by introducing a shock to the level of support in one or more PSE categories. The model then is solved by finding a new economic equilibrium post-shock. Results have a comparative static interpretation. The model can be run for any year in the 1986-2012 PEM database, or all of them at once. When run over all years, the model reports the policy impetus on various economic elements of the model for each year – it is not a dynamic simulation of how markets would unfold over the study

Box 4.1. **The OECD Policy Evaluation Model** *(cont.)*

period. This view allows for an estimation of the medium term impact of the policy in each year and to track how that medium-term impact changes over time as the policy set and underlying economic situation changes.

Policies are implemented in PEM according to the PSE classification. Each of the main kinds of support defined in this classification appears in the model with a specific initial incidence on producer and consumer incentive prices. The aim is to represent the incidence of support measures in the model in the same way as is implied by the classification of support measures for the PSEs. So, payments based on variable input use appear as a wedge between the supply and demand price of inputs, payments based on land between supply and demand price for land, and so on. Payments not requiring production or other payments with broad eligibility are modelled with price wedges designed to cause a uniform inflation in supply prices. That is, such policies do not affect the relative choice between eligible commodities or land uses.

As a part of the project on the Evaluation of Agricultural Policy Reforms in Switzerland, the Swiss module of the model is updated to represent three different geographical regions (plain, hilly and mountain) and to generate selected environmental performance indicators. As before, output markets and markets for purchased inputs are fully integrated within Switzerland. But the new Swiss module assumes that certain input markets are region specific such as land, cows and other farm owned capital to represent the production structure in each region. The regional representation and environmental assessment in the updated Swiss country module of the OECD PEM is documented in the technical document [TAD/CA/APM/WP(2014)28FINAL].

For general information on the PEM, see "Long-Term Trends in Policy Performance" (Martini 2011).

Periods of policy reforms

The assessment of two decades of agricultural policy reforms is structured along a discussion of five distinct periods that correspond to the sequential policy reform steps in Switzerland. The periods can be characterised in terms of the main policy changes as follows.[2]

1988-92: Pre-reform period

- Market price support accounted for approximately 90% of producer support
- Payments based on input use, and area and headage payments to specific products

1993-98: First policy reform (Law on Agriculture, Art.31a and 31b)

- The system of quantitative trade restriction was replaced by tariff rate quotas
- Guaranteed prices and production controls began to be deregulated
- Introduction of area payments based on current production area resulting in increased share of payments in producer support
- The level of PSE declined modestly

1999-2003: Second policy reform (AP2002)

- All state guarantees for prices and fixed processing margins were abolished

- Reform of the direct payment system, shift of a large part of direct payments to non-current area payments
- Introduction of the environmental cross compliance
- Significant decline in the level of PSE

2004-07: Third policy reform (AP2007)

- Start of phasing out of the milk quota system
- No major policy change in direct payments
- Almost constant PSE (a drop in 2007 reflecting price increases in world markets)

2008-12: Fourth policy reform (AP2011)

- Abolition of milk quota (2009)
- Elimination of export subsidies for primary agricultural products (2009)
- Increase in the payments based on ruminant animal numbers
- Modest decline in PSE due mostly to lower market price support for milk

Figures 4.1 and 4.2 illustrate how those policy changes translate into the Producer Support estimate as it is captured in the database that feeds into the OECD PEM, including the estimates for the three regions that the Swiss module distinguishes.[3]

The economic and environmental impact of those policy changes is assessed sequentially by posing the hypothetical "what-if" question: "What would have been the outcomes in that period if the policy changes had not taken place and instead the policy set of the previous period had stayed in place?"

For instance, the impact of the first policy reform in 1993-98 (RP 93-98) is assessed relative to the reference policies in the pre-reform period of 1988-92.[4] Similarly, the impact of the fourth policy reform in 2008-12 (RP 08-12) is evaluated by comparing it with the policy in

Figure 4.1. **Evolution of Producer Support Estimate modelled in OECD Policy Evaluation Model**

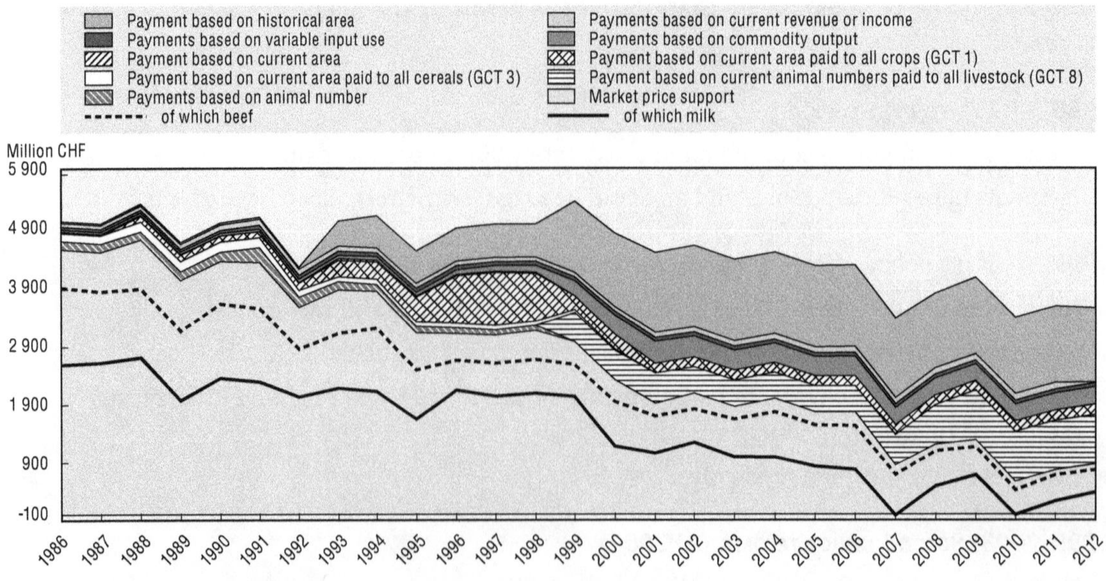

Source: Model simulation with OECD Policy Evaluation Model.

Figure 4.2. **Evolution of payments modelled in OECD Policy Evaluation Model by region**

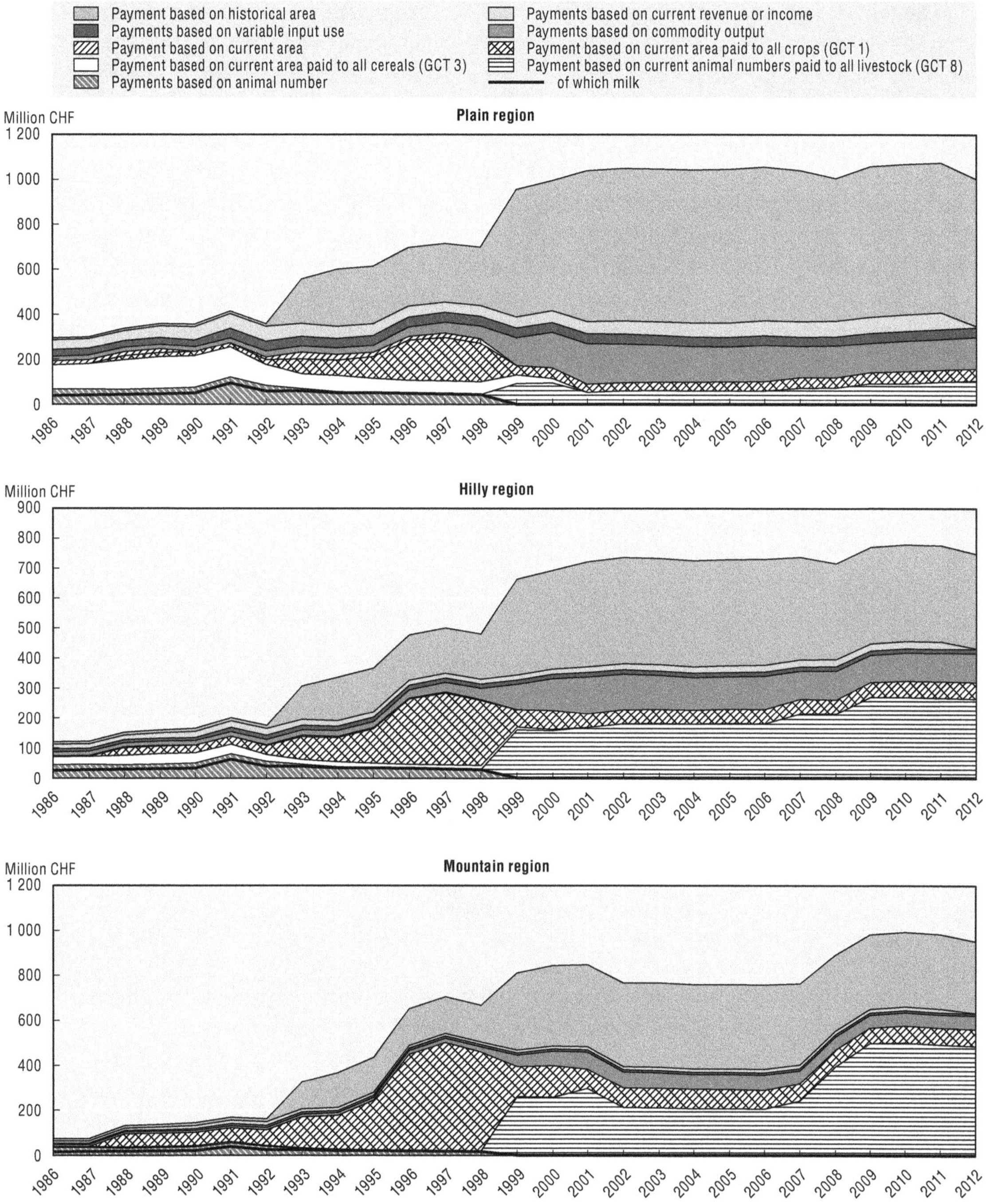

Source: Model simulation with OECD Policy Evaluation Model.

place during 2004-07 (RP 04-07) as a reference. The reference policies are constructed as an "average policy" during the reference period to smoothen the impact of exogenous factors, such as the impact of world prices and the exchange rate on the level of market price support. The model simulation generates the impact of the policy changes implemented for each year during the reform period, but the results are presented as average impacts during the reform period to isolate the policy impacts from other influences on the sector.

Economic impact of successive agricultural policy reforms

The reduction of market price support for arable crops which was partially cushioned by direct payments has consistently generated a policy incentive to lower production levels, particularly for wheat and coarse grains (Figure 4.3). The reform in 1993-98 introduced additional payments based on the current area of production and provided an incentive to put more land into crop production. Crop production became more extensive, using more area at lower yields. The extensive crop production improved the environmental performance of agriculture by reducing the total consumption of pesticide and inorganic fertilizer. However, the subsequent transformation of the direct payment system to payments based on non-current area and in conjunction with the expansion of payments based on animal number generated an incentive to shift land from arable crop production to pasture grass land in the next three reform periods.[5, 6]

Figure 4.3. **Simulated impacts of four policy reform programmes on arable crop production**

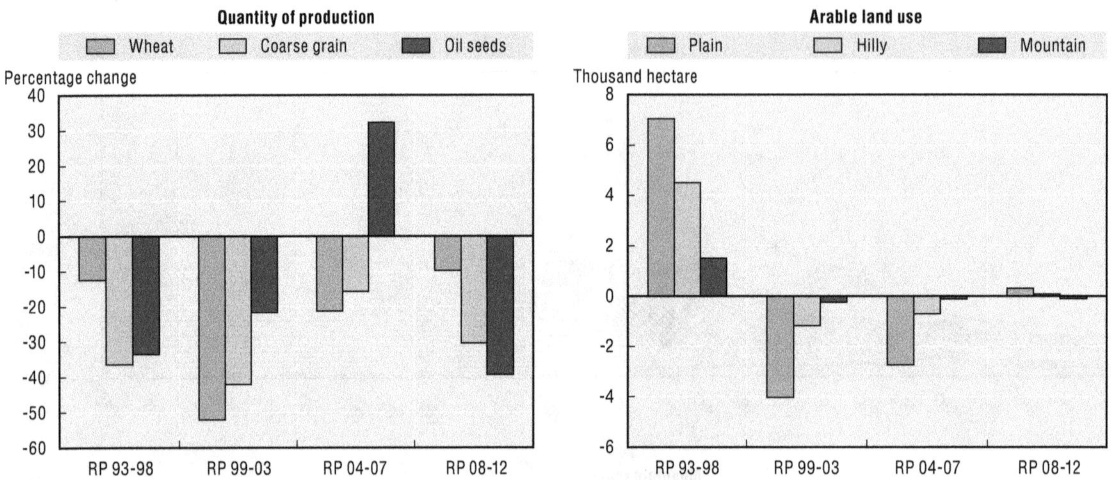

1. The impacts of policy reform are simulated relative to the reference (average) policy in the previous period. For example, the impact of RP 08-12 is simulated relative to a reference policy in RP 04-07.

Source: Model simulation with OECD Policy Evaluation Model.

Livestock production, which has long been subject to heavy-handed market intervention, responded to policy changes as well (Figure 4.4). The milk production quota system was in place until its phasing out was completed in 2009. The abolition of state guaranteed prices for milk in 1999 led to reduced milk production incentive and increased relative returns to beef production, which received a stable level of market price support. Phasing out of the milk quota system and the increase in payments based on animal numbers from 2004 onward increased the incentives to expand milk production, but to lesser extent than beef production. Expansion of milk production was limited as a consequence of the quota system that constrained milk output until 2009, as well as of the lower elasticity of milk demand. Policy changes clearly worked towards expanding beef cattle production, especially in hilly and mountainous regions, and increasing the stocking density (cattle per hectare of pasture).

The shifts in the volume and structure of production have affected the environmental performance of agriculture. Reduction of arable crops implies lower use of chemical inputs in that sector, while expansion of livestock production has led to increased environmental

Figure 4.4. **Impacts of four policy reform programmes on livestock production**

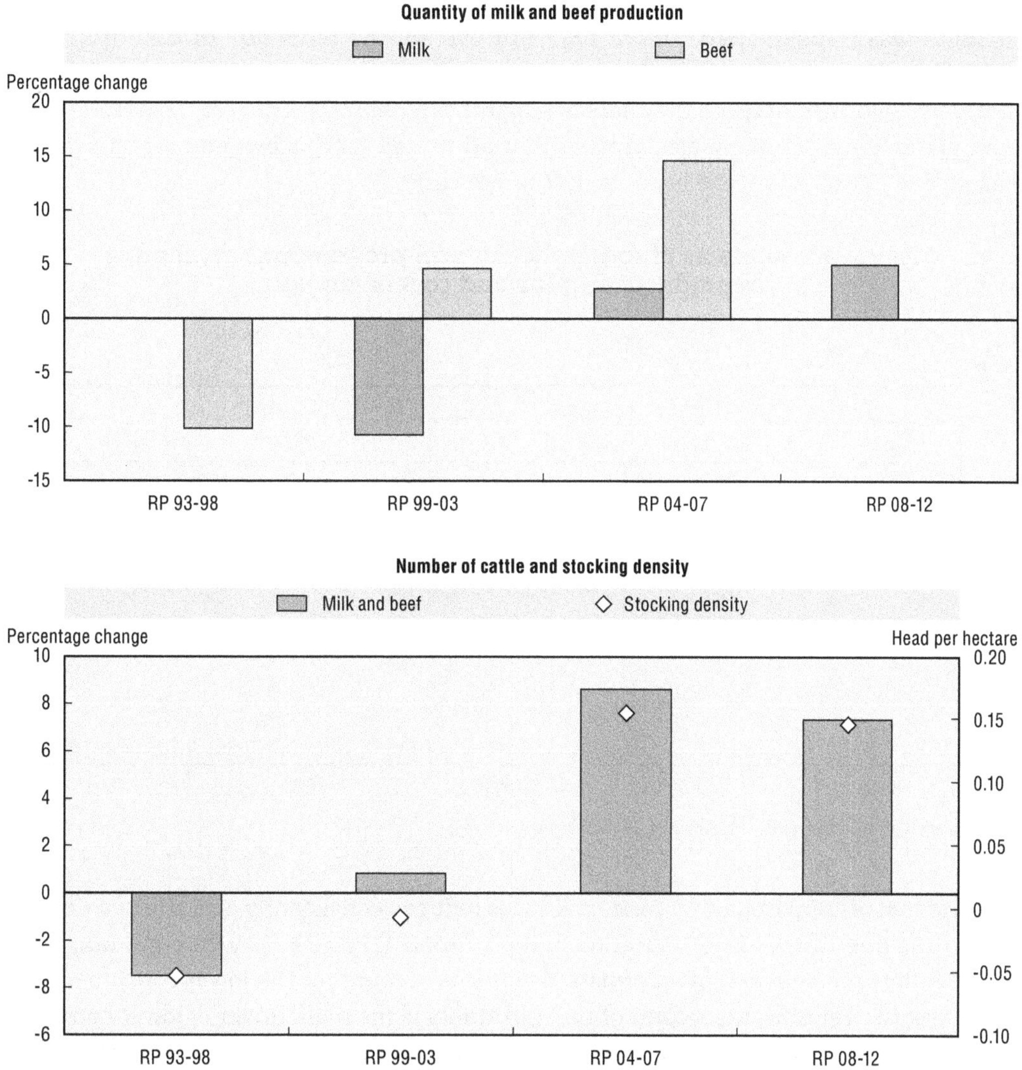

Source: Model simulation with OECD Policy Evaluation Model.

pressure stemming from animal husbandry. Environmental impacts of policy reforms are discussed in detail in the next section.

The policy reforms have had economic consequences for producers, consumers and taxpayers. Impacts on producers are expressed as changes in producer surplus and measured as returns above opportunity costs for farm owned factors: farmland, milk quota and the aggregate other farm owned capital.[7] The financial burden of support falls on consumers and taxpayers. Consumer impacts are measured in terms of change in consumer surplus, and costs to tax payers are calculated as the change in government expenditures stemming from direct payments, export subsidy expenditures and tariff revenues. In addition to the policy simulation scenario assessing the impact of policy reform programme relative to the policy in previous period, policy simulation to assess the welfare impact of the most recent reform programmes (RP 08-12) relative to the pre-reform period in 1998-92 is conducted to assess the overall reform impacts. The simulated impacts are presented as "Total" in Figure 4.5, 4.6 and 4.7).[8]

The policy reform programmes have reduced producer surplus except for the first policy reform in which the re-instrumentation of market price support with direct payment was implemented (Figure 4.5).[9] But overall, the reduction of cost of support exceeds the loss of producer surplus with an exception of RP 04-07. This indicates that the successive reforms improved the share of support captured by producers. This result holds even under the alternative assumption that all rented land is owned by non-farmers, leaking out 30% to 40% of the benefits accruing to land.

Figure 4.5. **Impacts of four policy reform programmes on change in producer surplus and cost of support**

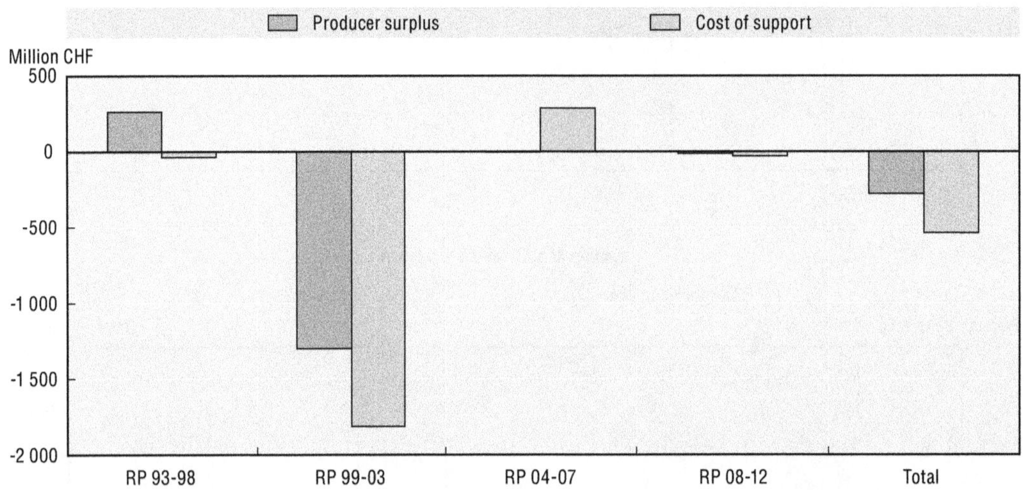

Source: Model simulation with OECD Policy Evaluation Model.

The cost of support paid by consumers was reduced consistently over the two decades of reforms due to lowering of market price support (Figure 4.6). While gradual trade liberalization and abolition of administrative prices account for the lower consumer costs in the first two reforms, the reform of the milk quota is the main driver of lower consumer costs in the later periods. On the other hand, the cost for taxpayers increased in three out of four reform programmes due to the partial replacement of market price support with direct payments.[10]

The reforms had differential impacts on farmers in different geographical regions (Figure 4.7). Payments targeted to the geographically less favourable mountain region redistributed economic surplus from producers in plain regions to those in mountain regions. All the policy reform programmes led to net loss for farmers in plain regions and producers in mountain regions experienced a gain from all the policy reform programmes except for the second policy reform programme (RP 99-03). Net surplus of producers in hilly regions was less affected by the policy reforms than other regions except for the RP 99-03.

The erosion of milk quota rents, due to phasing out of the milk quota during 2004-09, accounts for almost all of the loss in producer surplus in the last two reform programmes (RP 04-07 and RP 08-11) in all three regions. On the other hand, most of the producer gains are accounted for by the increase in surplus in land markets, in particular during the first phases of reform. The introduction of direct payments linked to land holdings increased the returns from land. In the mountain region, the gain for producers in the last two reform

Figure 4.6. **Impacts of four policy reform programmes on the changes in the sources of support**

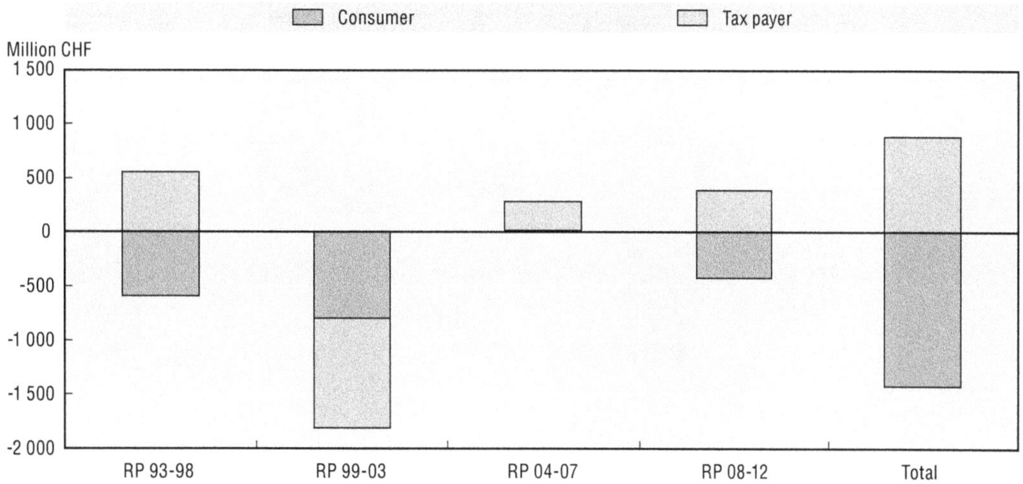

Source: Model simulation with OECD Policy Evaluation Model.

programmes stems largely from the appreciation of farm owned capital, in particular the value of farm-owned cattle. The opportunity cost of cattle assets was lifted by higher payments based on the number of animals. A change that was introduced in the last two reform programmes to compensate for the milk quota reform.

Environmental assessment of past policy reforms

Improving the environmental performance of agriculture has been a central element of policy reforms in Switzerland, and the country has made progress in reducing the environmental pressures stemming from agriculture (Chapter 2). Past reforms have, as discussed above, brought about important changes in the structure of production. Those changes also led to noticeable changes in environmental performance.[11]

Farm input use

Given the policy-induced reduction in arable crops, all four reform programmes have led to reduced consumption of chemical inputs, inorganic fertilizer and other variable inputs. Most of the reduction is registered in the plain regions, as this is the main crops producing area.

The steepest fall in pesticide use is attributable to the first decade of reforms, as this was the period with the sharpest re-orientation of policies (Figure 4.8). The RP 93-98 is estimated to have contributed by as much as -23% to the decline in pesticide use in plain regions, from 0.67 ktons of active ingredients to 0.51 ktons, and in the hilly region the policy impact reached -20% from 0.11 to 0.09 ktons. The next period of reforms under RP 99-03 led to an even more impressive reduction of 45% in both regions.

The effect of policy changes on the rate of pesticide reduction is somewhat tapering off in subsequent years, but still shows non-negligible impacts (See Annexe tables for all results).

The results for fertilizer quantities used in the main crops are similar to those of pesticides, with a continuous reduction of consumption of nitrogen and phosphorus compound, and the steepest drops induced by early policy reform efforts. In the first

Figure 4.7. **Impacts of four policy reform programmes on producer surplus by region**

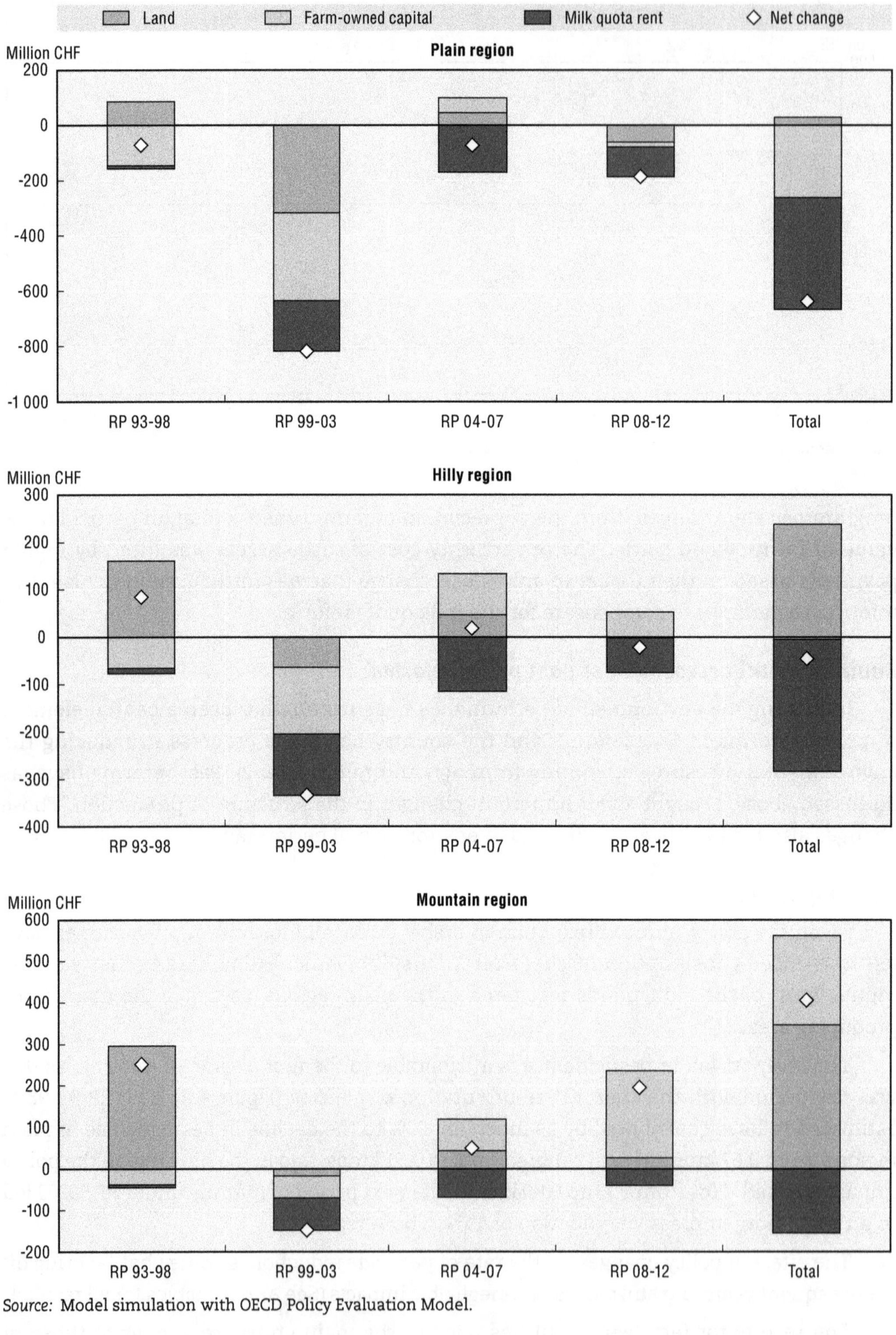

Source: Model simulation with OECD Policy Evaluation Model.

Figure 4.8. **Impact of four policy reform programmes on farm input use**

Change in ktons

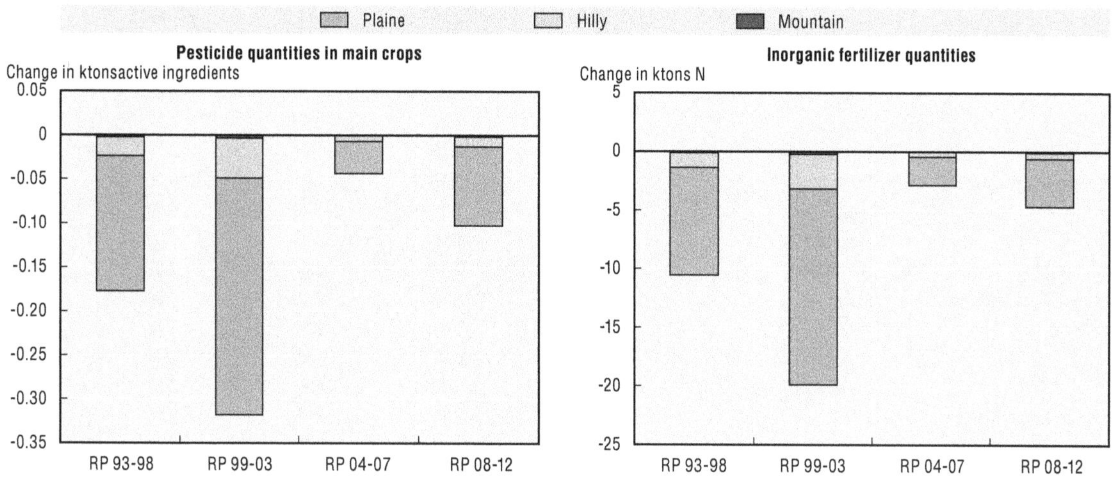

Source: Model simulation with OECD Policy Evaluation Model.

reform period, policy changes are estimated to have led to a decline of fertilizer use in main crops by -23% from 43 to 33 ktons of nitrogen in the plain region, and by a further -46% under RP 99-03.

Because policy shocks have resulted in only limited resource reallocation to alternative crops (e.g. sugar beet, potato, forage from intensive pastures), reforms had a small impact on the consumption of inorganic fertilizers on other land.

Manure from livestock

Changes in manure from cattle are directly linked to changes in the animal herd and to some extent to dairy productivity through increased nitrogen excretions in mature dairy cattle.[12]

In RP 93-98 and RP 99-04, the production of manure from cattle decreased, following the policy-induced shrinking of the animal herd, and subsequent policy changes are found to have led to an increase of manure production (Table 4.1). This evolution at aggregate level conceals changes at sub-sector and regional levels. Especially in RP 99-03, a sharp reduction of manure in mature dairy cattle was partially offset by an increase from beef cattle (e.g. herd of suckler cows and milk-fed veals).

The reduction of market price support for milk in RP 99-03 resulted in a reduction of the number of dairy cows that translated to a significant decrease of manure excretions. This policy reform is estimated to have contributed to a -10% decline of manure from dairy cattle, from 80 ktons (no-reform) to 72 ktons of nitrogen, with most of the reduction occurring in plain and hilly regions.

Subsequent policy changes increased direct payments to ruminant animals and phased out dairy quota. Those measures combined provided incentives to enlarge the dairy herd, which led to an increase in manure production. The largest increase of +14% (or +1.9 ktons) is found in mountain regions with RP 08-12.

The results for beef cattle are different. The consequence of the first reform period (RP 93-98) was a reduction of the beef herd and a drop in manure excreted, from 37 to 35 ktons

Table 4.1. **Impact of four policy reform programs on nitrogen budget**
Change in ktons N

		RP 93-98	RP 99-03	RP 04-07	RP 08-12
Manure from dairy cattle	Plain	0.21	-3.79	0.59	0.86
	Hilly	-0.18	-2.88	0.79	0.98
	Mountain	-0.10	-1.69	0.59	1.92
	Switzerland	-0.07	-8.35	1.97	3.76
Manure from beef cattle	Plain	-1.37	0.67	1.61	0.10
	Hilly	-0.60	1.23	1.48	0.78
	Mountain	-0.25	1.16	0.88	1.93
	Switzerland	-2.22	3.07	3.97	2.82
Manure from livestock	Plain	-1.17	-3.12	2.19	0.96
	Hilly	-0.78	-1.65	2.27	1.76
	Mountain	-0.34	-0.52	1.48	3.85
	Switzerland	-2.29	-5.29	5.94	6.57
Gross nitrogen balance	Plain	-4.99	-7.81	2.38	0.55
	Hilly	-1.02	-2.42	2.27	1.77
	Mountain	-0.31	-0.59	1.49	3.83
	Switzerland	-6.32	-10.82	6.13	6.16

Source: Model simulation with OECD Policy Evaluation Model.

of nitrogen. The reduction is of the same magnitude in all regions, around -6%. The subsequent three reforms fostered an increase of the beef herd and therefore implied more manure excretions. With the last reform (RP 08-12), manure from beef was almost stable in plain and slightly increasing in hilly region. Most of the increase occurred in the mountains (+1.9 ktons of nitrogen, or a 21% increase), which is explained by the reallocation of direct payments to those production areas.

Nitrogen intensity

The first two reforms resulted in a decrease of the overall nitrogen intensity in the country, thanks to a simultaneous reduction in inorganic fertilizer consumption in the arable sector and a decrease of manure production in livestock. At country level, the largest reduction occurred in RP 99-03. Policies in RP 04-07 led to an increase in manure production and slightly drove up nitrogen intensity in regions other than the plains. Policies in RP 08-12 fostered the increase in nitrogen emissions in the mountains (+ 3.8 ktons). This change is almost equivalent to the reduction in nitrogen inputs in plains, driven by lower fertilizer use, so that the total nitrogen inputs in Switzerland remained unaffected by the last reform.

Nitrogen balance

The Gross Nitrogen Balance (GNB) is the difference between "inputs" going to agricultural soils and "outputs" leaving the system with crop harvest and residues removed from soil. The balance is computed according to OECD-EUROSTAT soil-surface methodology. [13]

The first two reforms had a positive effect on GNB, i.e. the nitrogen surplus was reduced (Table 4.1). The contribution of policy changes to the improved nitrogen balance is estimated to be a -6% decline from 111 ktons (no-reform) to 105 ktons of nitrogen in RP 93-98, mainly due to reduced use of inorganic fertilizers. In RP 99-03, with the additional effect of a drop in manure from dairy cows, the balance was reduced by -11%.

Conversely, the two last reforms are estimated to have had deteriorating effect on GNB. Even if these reforms also reduced fertilizer use somewhat, they significantly encouraged the production of manure from beef and dairy cattle.

The direction and magnitude of impact per hectare in the first and third reforms are similar in all three regions, a reduction of 2-8% in the former, an increase of 6-9% in the latter (Table 4.2). For the other periods, the mountain region has a specific profile. The most striking specificity of mountains is the strong reaction to policies in RP 08-12, with the GNB going up by +24% from 55 kg nitrogen per ha (no-reform) to 68.

Table 4.2. **Impacts of four policy reform programmes on gross nitrogen balance in regions**

Kg N per ha

		RP 93-98	RP 99-03	RP 04-07	RP 08-12
Plain	No reform	126	107	79	84
	Actual	116	91	84	85
	Change	-10	-15	5	1
Hilly	No reform	100	96	79	84
	Actual	97	88	88	90
	Change	-4	-9	8	6
Mountain	No reform	65	65	59	55
	Actual	64	63	64	68
	Change	-1	-2	5	13

Source: Model simulation with OECD Policy Evaluation Model.

Phosphorus balance

The Gross Phosphorus Balance (GPB) is computed with the same methodology as GNB. Both balance show similar reactions to past policy changes. The GPB balance is found to respond to policy changes by a -4% decline from 11.9 ktons of phosphorus (no-reform) to 11.4 ktons in RP 93-98 and from 8.9 ktons to 7.6 ktons in RP 99-04 (-15%). Subsequent policy changes induced an increase, with the GPB going up for example from 5.8 to 7.0 ktons in RP 08-12 (+20%).

Greenhouse gas emissions

The main greenhouse gas emissions from the agriculture sector are methane (CH_4) from livestock and nitrous oxide (N_2O) coming from manure management and agricultural soils.[14] In Switzerland, methane emissions come exclusively from the animal sector, especially emissions from cattle enteric fermentation and manure management (88% of CH_4 in 2012). Therefore the size of dairy and beef herds are the main drivers of the volume of GHG emissions indicator. To some extent dairy productivity also contributes, as emissions are increasing with milk quantities produced per dairy cow.

Policies under RP 93-98 have had little impact on emissions from dairy cows (stable around 102 ktons of CH_4), but have reduced the emissions from the beef sub-sector, from 42 (no-reform) to 40 ktons of CH_4. The RP 99-04 has had a contrasting impact on dairy and beef sectors, the former decreasing their emissions by 10%, while the latter increasing by 10% (See Annex C).

Recent reforms are found to have led to rising methane emissions, induced by direct payments and removal of dairy quota that both led to larger livestock herds (Table 4.3). This

Table 4.3. **Impacts of four policy reform programmes on GHG emissions**

Change in ktons

		RP 93-98	RP 99-03	RP 04-07	RP 08-12
Methane	Plain	-1.3	-4.3	2.7	1.3
	Hilly	-0.9	-2.4	2.7	2.1
	Mountain	-0.4	-0.9	1.8	4.8
	Switzerland	-2.6	-7.5	7.2	8.3
Nitrous oxide	Plain	-0.3	-0.6	0.0	-0.1
	Hilly	-0.1	-0.1	0.1	0.0
	Mountain	0.0	0.0	0.0	0.1
	Switzerland	-0.4	-0.7	0.1	0.0
GWP (CO2eq)	Plain	-128.0	-271.2	56.1	-5.5
	Hilly	-40.6	-91.7	74.9	56.0
	Mountain	-13.5	-24.3	52.8	138.4
	Switzerland	-182.1	-387.1	183.9	188.8

Source: Model simulation with OECD Policy Evaluation Model.

is the case in both beef and dairy cattle and in all three Swiss regions, but more pronounced in mountain regions.

The sources of nitrous oxide emissions are agricultural soils and the management of manure in livestock. N_2O emissions from agricultural soils include direct soil emissions (i.e. application of inorganic fertilizers, manure, and crop residues returned to land) and indirect emissions (i.e., nitrogen from atmospheric deposition, leaching and run-off, excretions of animals on pasture range and paddock). Therefore, both animal and crop sectors are contributing to nitrous oxide emissions.

Not surprisingly, the profile of nitrous oxide emissions is similar to that of nitrogen inputs, with a sharp decrease of emissions occurring during the first reforms, followed by an increase in AP 04-07 and finally in RP 08-12 a stabilization of emissions.

The cumulated impact of methane and nitrous oxide emissions on climate change is traditionally assessed with the Global Warming Potential (GWP), computed from CO_2 equivalent emissions.[15] The Global Warming Potential declines with the first two reforms, from 5 950 ktons CO_2eq (no-reform) to 5 768 ktons in RP 93-98, and then from 5 870 ktons CO_2-eq to 5 484 in RP 99-03. More recent reforms had led to increases of the Global Warming Potential.

It is worth reminding that PEM does not include livestock other than cattle. Of particular economic importance in Switzerland are pigs (1.544 million head in 2012, including 0.128 million swine), poultry (9.978 million, including 2.520 million laying hens) and sheep (0.417 million heads) (OFAG, 2013). The two first are monogastric animals. Contrary to ruminants, they are not significantly contributing to methane emissions from enteric fermentation, but manure from swine is almost exclusively managed as liquid or slurry in Switzerland, yielding high emission rate of methane from manure management (Bretchler, 2011). Pigs and poultry are also contributing to nitrogen excretions and nitrous oxide emissions. In the current PEM model, excretions and emissions from these sectors are assumed to be fixed to their current baseline value. However, directly with the changes in support to livestock, and indirectly through the feed market and substitution in demand between commodities (i.e. red and white meat), policy reforms have an impact on intensive enclosed breeding production and their environmental outcome. Some non-OECD studies on environmental assessment of the Swiss policy reforms are characterised in Box 4.2.

Box 4.2. **Some non-OECD studies on environmental assessment of the Swiss policy reforms**

Spiess (2011) calculates the national nitrogen and phosphorus balances of Swiss agriculture from 1975 to 2008 by using a farm-gate balance approach according to OSPAR guidelines. In 2008 the farm-gate balance showed surplus of 108 kg N/ha and 5.5 kg P/ha for Swiss agriculture. These farm-gate balance figures are higher than those given by soil surface balance method used in OECD calculations. The farm-gate balance approach is more precise method than soil surface balance and usually gives higher nutrient surplus figures than soil surface balance method.

Herzog et al. (2008) evaluate the impact of environmental cross-compliance (Proof of Ecological Performance, PEP) on nitrogen and phosphorus pollution from Swiss agriculture. This a combined monitoring schemes with evaluation of cause-effect relationships in selected case-study areas. According to the study, reduction of mineral and manure nitrogen due to PEP requirements resulted in a decrease of mean nitrate leaching about 10 kg N per ha per year (from 49 to 39 kg N per ha per year). Use of cover crops resulted in a decrease of nitrate leaching of about 5 kg N per ha per year. Overall PEP requirements reduced nitrate leaching by 29% (16 kg per ha per year) for the arable land. Herzog et al. estimate that phosphorus pollution of surface waters from agriculture has decreased by 10-30% since the introduction of PEP requirements.

Aviron et al. (2009) evaluated the impact of Ecological Compensation Areas (ECAs) on farmland biodiversity (species richness) in three regions representative of the different farming types (arable, mixed arable-grassland, and grassland) in central Switzerland. The evaluation included both meadows and wildflower strips and the difference between ECA and conventionally managed fields was assessed for vascular plants, butterflies, ground beetles and spiders. ECA meadows contained a greater number of species of plants, ground beetles, and butterflies but not of spiders. ECA wildflower strips contained 8%-60% more species of plants, ground beetles, and spiders than on arable fields. However, very few of the plant and arthropod species observed were so called red-list "threatened" species while seven ground beetle and six butterfly species were listed in Red Data books. Thus, overall ECAs benefit more common species than threatened species.

Economic and environmental impact of forthcoming agricultural policies in 2014-17

This section assesses the possible or likely impacts of forthcoming agricultural policy changes between 2014 and 2017 (AP 2014-17). The main objective of this reform is to reduce the general area payment and target direct payments more to specific policy objective such as bio-diversity, landscape and other public benefits of agriculture as overviewed in Chapter 3.

Based on the available information, the forthcoming payments are classified according to the OECD PSE methodology, and subsequently modelled in the PEM framework. The market price support, payments based on output and payments based on variable input use during 2014-17 is assumed to be constant at 2012 level. Figure 4.9 presents the evolution of producer support estimates included in the PEM in three geographical regions (see the technical document [TAD/CA/APM/WP(2014)28FINAL] for the regional disaggregation by the classification of producer support). The level of total

Figure 4.9. **Evolution of Producer Support Estimates modelled in OECD PEM for AP 2014-17**

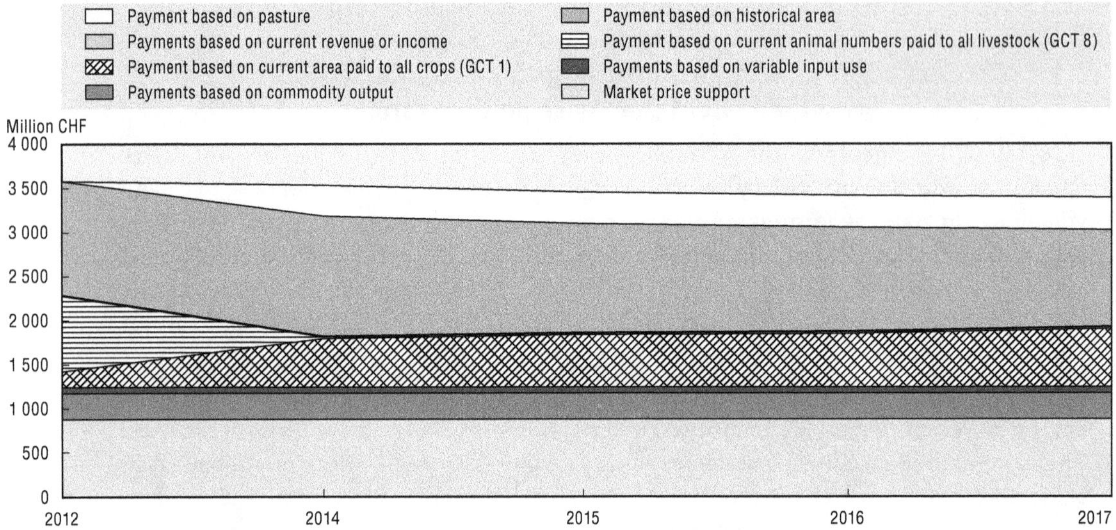

Source: Model simulation with OECD Policy Evaluation Model.

payments modelled in the PEM during 2014-17 is slightly lower than the 2012 level.[16] In the plain region, the payment based on non-current area (PSE classification of E) is partially replaced with current area payment to all crops. In the mountain area, the payments based on the area of pasture replaces the payments based on animal number and the payments based on non-current area.

The policy simulation in this section investigates the economic and environmental impact of policy change proposed in 2014-17, using 2012 as a base year. The results are presented as an average medium-term impact of forthcoming policies in 2014-17. The results show that the new policy reverses past trends to some extent: it is expected to reduce the size of livestock herds particularly in the hilly and mountain region but to lead to only small changes in crop production.

The production impact of forthcoming policies is most prominent in the livestock sector. The shift of direct payment from animal numbers to land area induces producers to use land more extensively and reduces the stocking density (Figure 4.10). Overall stocking density in Switzerland is expected to decline from 2.1 to 1.8 cattle per hectare of pasture. In particular, the decline in the stocking density is the most significant in the hilly region: 20% lower stocking density from 2.1 to 1.7 cattle per hectare of pasture, driven by the reduction in animal numbers. In the plain region, the direct payment system is to shift partially from non-current to current area payments for crop production. This reorganization shifts some of the pasture land use to crop production.

The policy shift from the payments based on animal number to land area in hilly and mountain regions induces less intensive livestock production in these regions, while the impacts of the new policies on arable crop production are negligible (on average 1.1% increase). This is consistent with the findings of the SILAS model developed by Agroscope Reckenholz-Tänikon, which expects only a small increase in the production of cereals (1.7%), when comparing the 2017 situation with and without the reform of direct payments (Zimmermann et al., 2011).

Figure 4.10. **Impact of AP 2014-17 on animal number and stocking density**

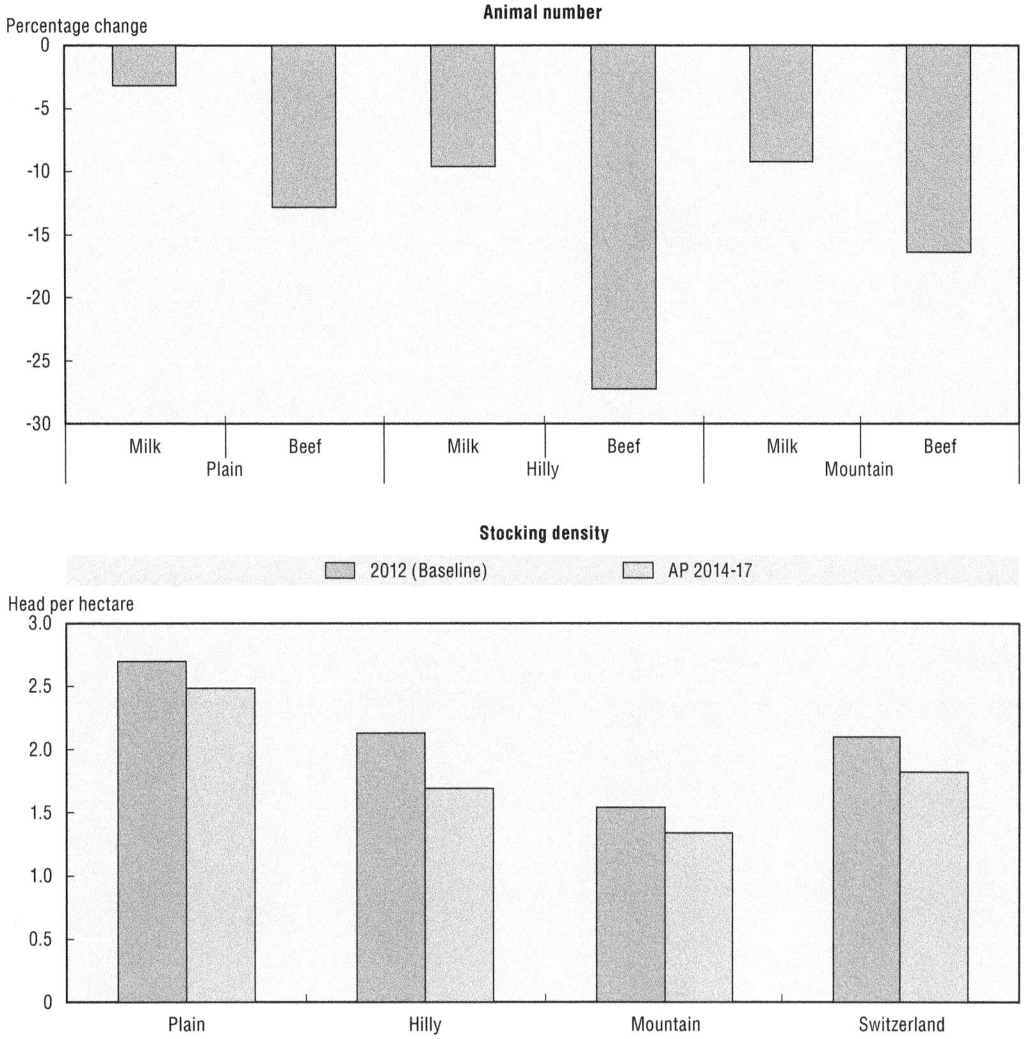

Source: Model simulation with OECD Policy Evaluation Model.

The PEM simulation shows the negative impacts of AP 2014-17 on both milk and beef production, which are equivalent to about 4.2% and 8.2% decline from 2012 level, respectively, with hilly regions accounting for the largest share in decline (Figure 4.11).[17] Again the findings are consistent with the simulations of Agroscope, which expect a reduction of 2.5% in milk production and 5.3% in beef (Zimmermann et al., 2011).

Most of the changes in environmental performance are expected to be driven by the animal sector, as production effects on crops are negligible. Change is concentrated in hilly and mountain regions. The plain region sees little change in nitrogen inputs, except a minor reduction of manure excretions from dairy cows and a small increase in beef cattle.

The reform is expected to generate an improvement of the nutrient balance. In relative as well as in absolute terms, the impact is stronger on the beef sector than on the dairy sector (Figure 4.12). On average during the four-year period, manure from beef cattle could be -18% lower than in 2012, from 38 to 31 ktons N. In dairy cattle, manure is expected to decrease by -6%, from 68 to 65 ktons N. Overall, the nitrogen surplus is expected to go down by -12% from 87 in 2012 to 76ktons N.

Figure 4.11. **Impacts of AP 2014-17 on the quantity of milk and beef production**

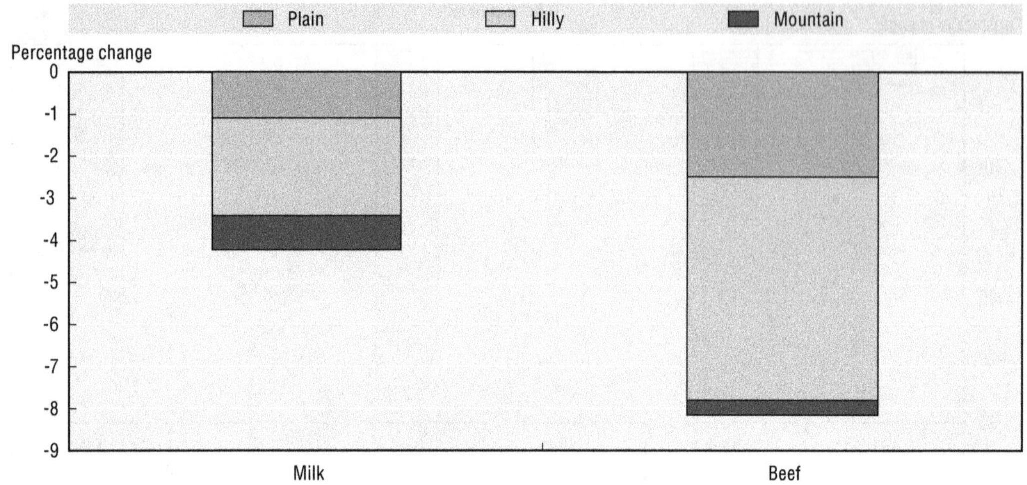

Source: Model simulation with OECD Policy Evaluation Model.

Figure 4.12. **Impact of AP 2014-17 on selected agri-environmental indicators**

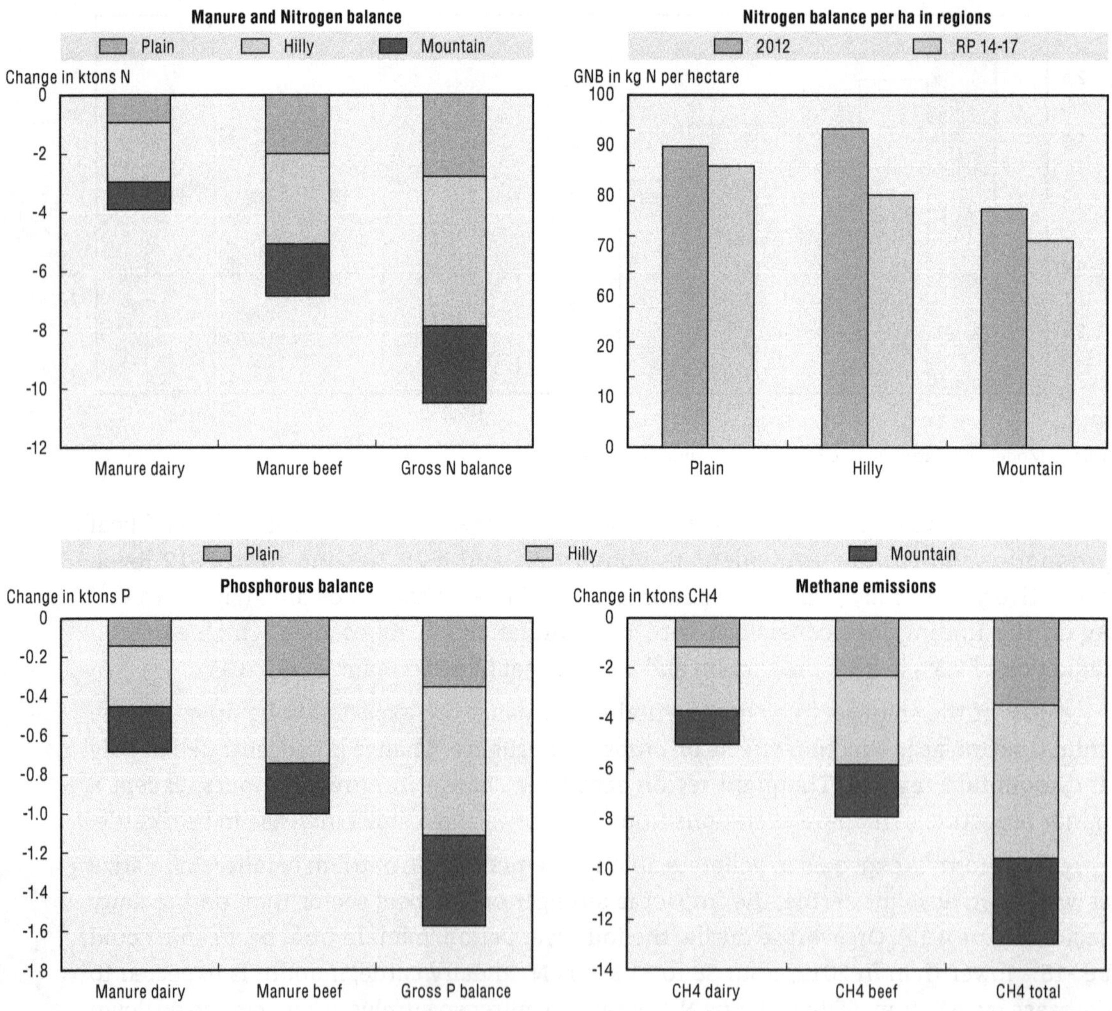

Source: Model simulation with OECD Policy Evaluation Model.

With 90 kg of nitrogen per hectare, the hilly region is the agricultural region with the highest nitrogen surplus in the 2012 baseline: higher than in the plain region (85 kg N) and much higher than in the mountain region (68 kg N). It is estimated that the new policy set will reduce surplus in hilly regions by 21% to 72 kg N. In the mountains, the surplus is estimated to go down by 13% to 59 kg per ha.

The estimates also point to a parallel reduction of phosphorus surplus from 6.8 to 5.2 ktons P for the whole country. The emissions of greenhouse gases could be reduced as well: the expected reduction of the animal herd would lead to 8% lower emissions of methane (from 160 to 147 ktons CH_4) and 5% lower emissions of nitrous oxide. In terms of global warming potential, this translates in a 377 ktons CO_2eq reduction, from 5 600 to 5 223 ktons CO_2eq. This reduction is in the order of magnitude of changes induced by reform AP 99-03 (-387 ktons CO_2eq).

Overall, the policy changes improve the transfer efficiency of the policy set. The simulation results show that the marginal decrease in producer surplus (CHF 12.6 million) is smaller than the CHF 189.5 million reduction of cost to taxpayers in AP 2014-17 on average (Figure 4.13).[18] Hence, this is an overall benefit to the society. The payment based on land (arable crop production, pasture or historical area) results in more pronounced gains in producer surplus than the payments based on animal numbers. This result is still robust under the alternative assumption regarding the share of benefits accruing to land which remains in the agricultural sector. Under the extreme assumption that all the rented land is owned by non-farmers, the decline in producer surplus can be as large as CHF 152.5 million.[19]

Figure 4.13. **Impacts of AP 2014-17 on producer surplus and tax payer's cost**

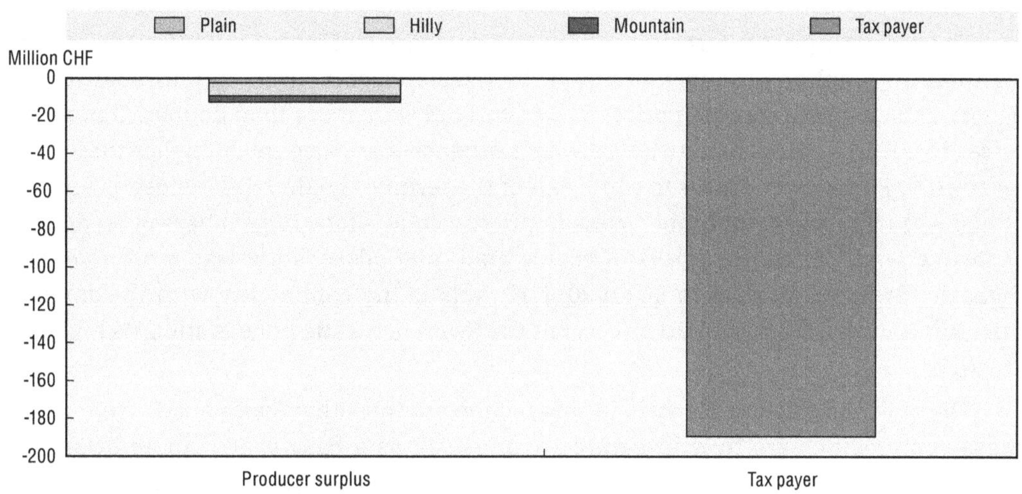

Source: Model simulation with OECD Policy Evaluation Model.

The composition of changes in the producer surplus in 2014-17 by region shows a marginal net loss of producer surplus in all three regions (Figure 4.14). The change in producer surplus shows that AP 2014-17 maintain the policy framework to target support to geographically disadvantaged areas. Figure 4.14 also shows that increasing land value accounts for all the gains for the producer. The reorganization of the direct payment system more to land area based payments reallocates producer surplus from farm-owned capital (mainly the value of cattle owned) to land, particularly in hilly and mountain regions.

Figure 4.14. **Impacts of AP 2014-17 on producer surplus by region**

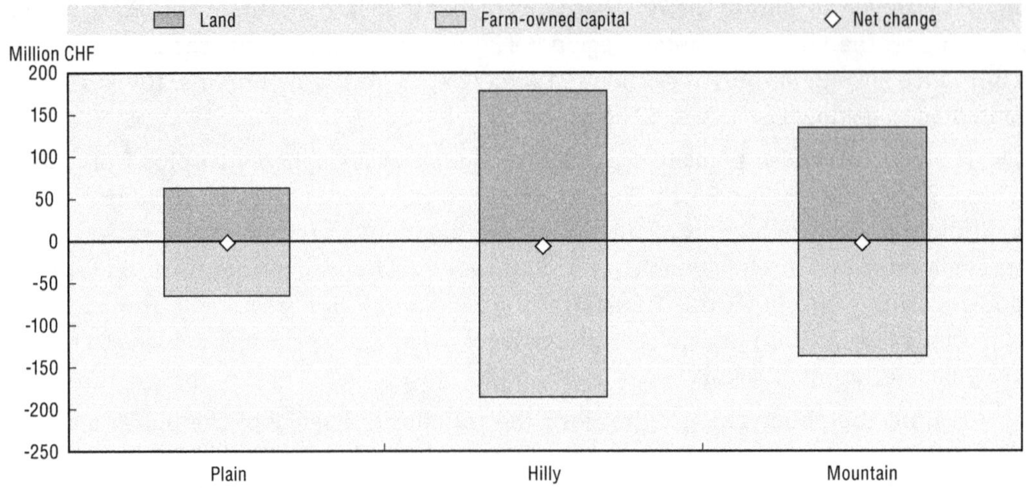

Source: Model simulation with OECD Policy Evaluation Model.

Economic and environmental impacts of further market integration with the European Union

This section assesses the impact of a hypothetical policy reform in the direction of further trade liberalization in the Swiss markets for agricultural commodities. More specifically, the section considers a scenario under which Swiss agricultural markets are integrated with EU markets so that producer prices in Switzerland are aligned with EU domestic price.

The producer price information in the PSE database (calibrated in PEM model) shows significantly higher producer prices in Switzerland compared to European Union (Figure 4.15). In particular, beef prices in Switzerland were more than double EU producer prices in 2010-12. Prices paid to Swiss wheat producers are also much higher than those received by EU farmers. This is to some extent the result of quality differences as produce of higher quality wheat for human consumption accounts for a larger share of production in Switzerland.[20] Producer prices of coarse grain and oilseeds are also more than 50% higher in Switzerland than in EU in 2010-12. Milk is the commodity with the smallest price gap between Swiss and EU prices, but the Swiss domestic price is still 27% higher in 2010-12.

The policy simulation scenario in this section assumes that Swiss and EU markets in these commodities are fully integrated, using 2012 as a base year.[21] In addition, the alternative scenario introduces complementary payments in the form of an increase in payment based on non-current area to investigate market and welfare effects of such compensatory payments. Under this alternative scenario the increase in non-current area payment is 35% (CHF 451 million), keeping the regional allocation share of this category of the payment constant. The results of some other quantitative studies on EU market integration are discussed in Box 4.3.

Due to lower domestic commodity prices, Swiss consumers are estimated to gain around CHF 1.70 billion (Figure 4.16). The total cost of supporting producers would be reduced by CHF 1.49 billion, taking into consideration the increase in consumer surplus,

Figure 4.15. **Producer prices in Switzerland and European Union in 2010-12**

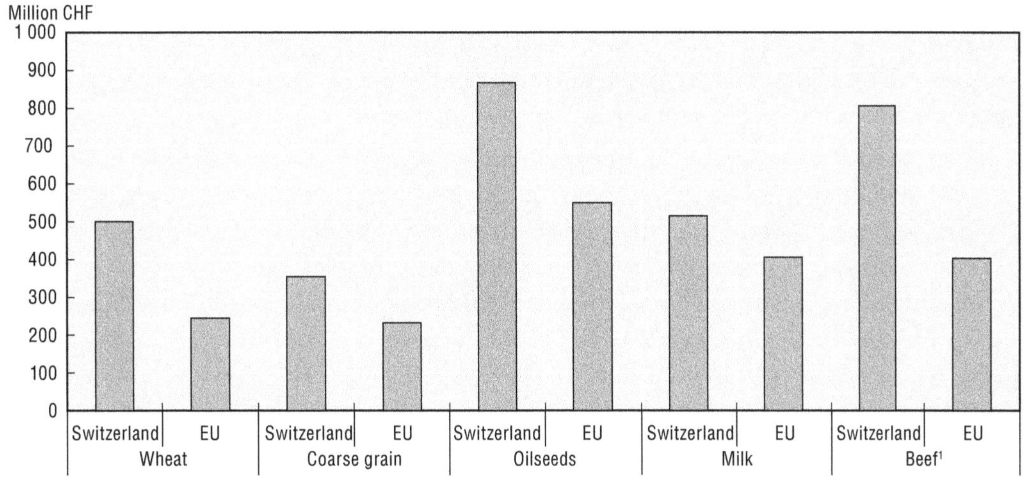

1. Beef price is expressed per 100 kg

Source: Calibration from OECD PSE/CSE database 2013.

Figure 4.16. **Impacts of EU market integration on producer, consumer and tax payer surplus**

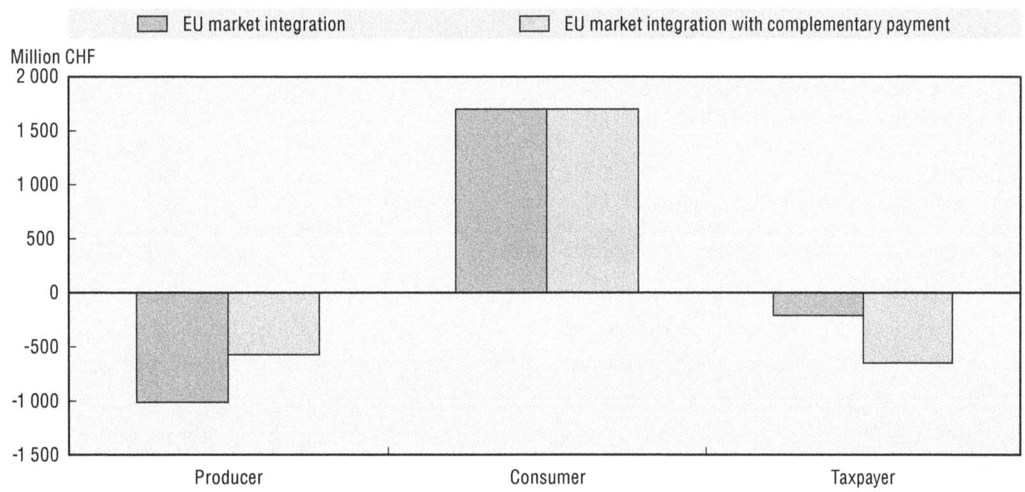

Source: Model simulation with OECD Policy Evaluation Model.

the decline in payments linked to current output or input and the loss of tariff revenue. On the other hand, the loss of producer surplus is estimated at CHF 1.01 billion, which is far less than the reduction in the cost of support.[22] In comparison with the impact of previous policy reform programmes, the size of loss in producer surplus is far less than that of RP 99-03 when all state guarantees for prices and fixed processing margins were abolished.

The complementary payment would limit the loss of producer surplus to CHF 573 million. In this case, the cost of producer support (an increase in consumer surplus and a decrease in tax payer surplus) would be still CHF 1.05 billion lower, which is much more than the loss of producer surplus. The simulation indicates that further trade liberalization with the European Union would improve the efficiency of producer support policies and increases the overall economic welfare with or without certain complementary payments.

An important additional effect that is not accounted for in this analysis is the potential of market integration to make Swiss agricultural sector more competitive through lower purchased input costs and more competition with foreign producers. Moreover, such reform would also improve the competitiveness of the Swiss downstream industries that would get access to cheaper primary agricultural inputs.

The production impacts of EU market integration would be largest for wheat. Without complementary payments, the production of wheat would be reduced by 12%, due to closing the current large price differentials of these commodities between Swiss and EU markets (Table 4.4). The downward estimates for wheat are most likely exaggerated as the model simulations assume a complete closing of the price gap between Switzerland and the European Union. This may not be fully realistic as Swiss wheat is typically of a higher quality, and fetches a higher price, than the average EU wheat. The liberalization increases the relative profitability of coarse grain, oilseed and other arable crops, shifting production to those commodities.[23] Milk production would shrink by 7%. Complementary payments would limit the reduction of domestic production to a certain extent under the EU market integration.

Table 4.4. **Impact of EU market integration on production and consumption**

	Wheat	Coarse grain	Oilseeds	Milk	Beef
% Change in quantity of production					
EU market integration	-12	9	22	-7	-9
EU market integration with complementary payment	-7	7	16	-7	-5
Share of domestic production in consumption					
Baseline (2012)	52	78	18	104	90
EU market integration	44	90	15	88	43
EU market integration with complementary payment	46	89	15	88	45

Source: Model simulation with OECD Policy Evaluation Model.

The lower domestic price increases the consumption of all commodities, inducing more imports. As a consequence, shares of domestic production in consumption would decline except for coarse grains. In particular, the self-sufficiency of beef would decline significantly.[24] Similarly, the share of domestic production in wheat consumption would decrease from 52% to 44%. Switzerland would become a net milk importer, but 88% of domestic consumption would still be satisfied by domestic production even under the condition of market integration with EU.[25]

The impacts of market integration with EU on animal numbers tend to be larger in the plain region than in hilly and mountain regions (Table 4.5). In hilly and mountain regions, milk and beef cattle numbers would decline only modestly because the size of payments targeted to geographically disadvantaged area is high so that the decline in market revenue would not change production incentive to a large extent. Consequently, the stocking density in hilly and mountain regions would change much less than in plain regions.

The production changes would directly translate into improved environmental performance of the agricultural sector, with the overall change mostly driven by the livestock sector.

Consumption of inorganic fertilizer is expected to decrease slightly (Figure 4.17). Manure excretions would go down significantly from 140 to 130 ktons of nitrogen (-7%). For

Table 4.5. Impact of EU market integration on animal number and stocking density

	Baseline (2012)	EU market integration	EU market integration with complementary payment
% Change in animal number			
Milk			
Switzerland		-6	-6
Plain		-8	-8
Hilly		-4	-3
Mountain		-6	-7
Beef			
Switzerland		-10	-6
Plain		-17	-14
Hilly		-6	-1
Mountain		-3	3
Stocking density (cattle per ha)			
Switzerland	2.1	2.0	2.0
Plain	2.7	2.4	2.4
Hilly	2.1	2.1	2.2
Mountain	1.5	1.6	1.5

Source: Model simulation with OECD Policy Evaluation Model.

Figure 4.17. Impact of further trade liberalization on selected agri-environmental indicators

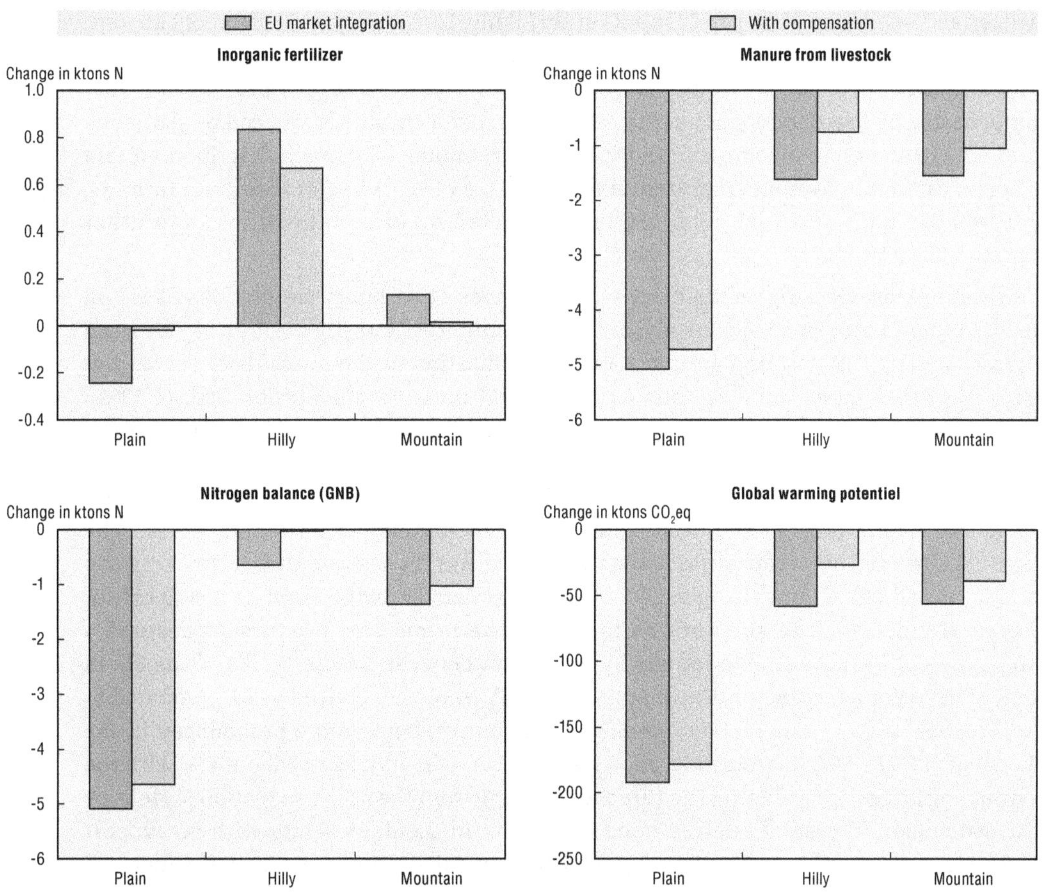

Source: Model simulation with OECD Policy Evaluation Model.

the dairy sector, the magnitude of reduction is similar in all three regions, while manure reduction from beef cattle is much stronger in hilly and mountain regions than in plain region. The consequence of further market integration with the EU would be an improvement of the GNB, with a reduction of 10 ktons of nitrogen surplus in Switzerland.

The nutrient surplus per hectare would be reduced by 6 kg of nitrogen in plain and hilly regions, and up to 10 kg in the mountains. The same improvement of the nutrient balance holds for phosphorous. Finally, it should be noted that the Global Warming Potential would be declining by 7%, from 5 600 to 5 222 ktons of CO_2eq.

Box 4.3. **Some other quantitative studies on EU market integration**

There are other studies simulating the quantitative impacts of EU market integration on Swiss agriculture based on equilibrium models. However, there is a number of differences in the model assumptions between those and the OECD PEM used in this report. For example, the PEM does not incorporate any external projection in market developments or structural changes in the farm sector. As such those simulation results cannot be strictly compared to those carried out by the OECD PEM. Here we consider some of the most relevant studies found in the literature.

A study with the SILAS model (Agroscope) was conducted to assess the impact, on agricultural production and income in 2016, of a free trade agreement between Switzerland and the EU (Confédération Suisse, 2008). The assumptions are that with such an agreement, commodity prices converge to those of the four neighbouring countries of Switzerland, while the price gap of inputs is shrinking. Direct payments are kept constant at the level under AP 2011. The study finds that milk production would increase, whereas crop and meat production would decrease. Lower commodity receipts are not fully compensated by lower production costs, so that the sector income is reduced by about one third. The reduction in income is found to be more pronounced in plain than in mountain regions , as direct payments represent a higher share of revenues in mountain farming, and because milk and veal meat are less impacted by price reductions than other commodities (Confédération Suisse, 2008).

A recent report focusing on the dairy sector provides other policy simulations based on the CAPRI market model (Bonn University) and the supply model SWISSland (Agroscope).The FTA scenario assumes a full liberalisation of the Swiss dairy sector, but not of the other agricultural sectors, e.g. crops and meat (Confédération Suisse, 2014). According to CAPRI model, liberalisation of the dairy sector with no additional compensatory measures (scenario "S_0") would result in a reduction of 40% in the domestic price of butter and cream, and 30% for whole milk powder. The price of cheese would slightly increase (+3%) and the price of skimmed milk powder remain stable. The slight increase in the producer price for cheese is driven by the removal in the scenario "S_0" of the so called processing aid for cheese that acts in Switzerland as a support for cheese production. The simulations to assess the impact on production assume a progressive 20% reduction in price gap in farm input between Switzerland and the EU. In 2025, after sector adjustments, the quantities of milk would be reduced by 6% compared to the situation without any market opening, the reduction being more pronounced in the plain regions. The SWISSland model shows a reduction of dairy cows of about 4%. With the contracting prices and volumes, the turnover deriving from the milk production shrinks by CHF 640 million in case of full free trade agreement in the dairy sector with no support measures, reducing the profits of milk producers and dairy industries. Given the increase

Box 4.3. **Some other quantitative studies on EU market integration** *(cont.)*

in consumer surplus, the overall impact of liberalisation on welfare is estimated to be positive (CHF 176 million). The analysis shows that the reduction of the income of milk producers can be strongly alleviated if support measures within the current limits continue to be paid to the milk sector.

Two studies by ETH Zürich make use of the agricultural sector model S_INTAGRAL, aiming at assessing the impact of different producer price scenarios on the land use, livestock and greenhouse gas emissions until 2020. Peter et al. (2009) shows that the greenhouse gas emissions could be about 4% lower under a free trade agreement with the EU. However, longer term structural adjustments in the agricultural sector (by 2020) would lead to a rebound in emission, stabilising around their baseline levels. More detailed analysis shows an increase in the emissions of CH4 from the enteric fermentation of herbivores under the liberalisation scenario (in comparison to the status-quo) which is offset by lower CH4 emissions from manure management, as well as lower N2O emissions from agricultural soils (Hartmann et al., 2009, p.19). According to these studies, the number of cows and total cattle would be fairly close in both scenarios, after sector adjustments to liberalisation.

Notes

1. The plain, hilly and mountain regions are defined according to agricultural regions defined by the Swiss Federal Statistical Office.
2. See Hofer, E. (2009) "A Survey of Swiss Agricultural Policy Reform (1982-2007)" Federal Office of Agriculture for the definition of agricultural policy reform stages in Switzerland. For simplicity, the four reform periods in Chapter 3 are referred to according to the years of their implementation: RP 93-98, RP 99-03, RP 04-07 and RP 08-12.
3. Note that not all categories of the PSE are included in the OECD PEM. The model covers only five aggregate commodities (wheat, coarse grain, oilseeds, milk and beef) and excludes payments based on non-commodity criteria, payment based on variable input use with input constraints, and certain payments based on current area/animal number whose commodity or commodity group are not covered by the PEM. Although the payments based on animal numbers of all livestock (GCT7) are not covered by the PEM, the payments for "holding of livestock under difficult conditions" and "regularly keeping animals outdoors" are included, assuming that the payments are only based on the number of cattle. Similarly, the payments based on the current area of production of grain and oilseeds (GCT 10) and all crops except wine (GCT11) are assumed to be paid based on the area of production of all crops (GCT1).
4. In this case, PEM policy simulation replaced factor subsidy rates of the policies in each year of the baseline reform period (RP 93-98) with an average factor subsidy rate during the previous period (1988-92). The impacts of RP 93-98 are presented as average impacts of simulations conducted each year between 1993 and 1998.
5. The PEM simulates the outcomes as if the policy changes had not taken place and instead the policy set of the previous period had stayed in place without taking into account any exogenous changes other than the modelled policies. The parameters of the model are calibrated to replicate medium-term (approximately 5 years) equilibrium of the markets. Therefore, the simulated impacts do not necessarily match the observed changes in production, land use, animal number and environmental performances.
6. Moreover, several limitations of the PEM have to be recognized in in interpreting the simulated outcomes. First, the PEM models only five aggregate commodities (wheat, coarse grain, oilseeds, milk and beef) and a sub set of payments recoded in PSE data. Second, the simulated outcomes are based on the assumptions of supply and demand relations, with input and output markets approximated with constant elasticity linear equations. Third, it does not explicitly model regulatory measures in place or cross-compliance conditions attached to the payment eligibility (e.g. Proof of Ecological Performance). See Box 3.1 and the technical document *www2.oecd. org/oecdinfo/info.aspx?app=OLIScoteEN&Ref=TAD/CA/APM/WP(2014)28* for more information on the OECD PEM.

7. The OECD PEM does not differentiate between farm-owned and rented land. The simulation results assume that land is fully owned by farmers. However, to determine who benefits from the programme value accruing to land, average rental rates can be used to disaggregate this into farmers who own their land and landowners who rent to farmers. The data provided by Agroscope based on the Swiss FADN survey shows that the share of land that is rented is stable at 41%-45% in the plain region during 1990-2012. The share of rented land in hilly and mountain region is lower: 35%-40% in the hilly region and 33-41% in the mountain region. That is, 60%-70% of the increase in producer surplus to land could accrue to farmers, and the rest to landowners who generally do not farm. This can be considered as a lower bound, as it is likely that some rental arrangements are between farmers, such that the land owner and renter are both farmers, and hence any surplus stays in the farming sector.
8. "Total" represents the welfare impact of RP 08-12 taking the policies in the pre-reform period as reference policies. Therefore, it is not necessarily equal to a sum of the impacts of four policy reform programmes whose reference policies are those in previous reform period.
9. The OECD PEM does not explicitly model structural change through farm exit and entry. Lower producer surplus may lead to the exit of economically non-viable farms. In reality, the number of farms declined 1.8% per year in 2000-12. The rate of decline in producer surplus per farm is most likely less than that at the sector level.
10. Lower cost of support paid by consumers can also be interpreted as an increase in consumer surplus. Similarly, higher cost of support financed by taxpayers indicates reduction in taxpayer's welfare.
11. The study with the PEM model does not asses the direct impacts of policy reforms on the conservation of biodiversity and on landscape quality and diversity, which are also important objectives of the Swiss policy. The impact on biodiversity will be discussed in the final report, making use of published studies.
12. For the animal categories not modelled in factor markets (e.g. swine, pigs, poultry, sheep) constant excretions for each region are assumed.
13. The same methodology and coefficients are used in this study and by the Federal Office of Statistics of Switzerland (OFS) for the computation of nutrient balances in the OECD compendium of agri-environmental indicators (OECD, 2013a). However, the baseline values in this study are different, as it considers the three officially defined agricultural regions, plain, hilly and mountain, whereas OFS also includes in the GNB land classified as non-agricultural (alpine summer pastures) (Kohler, 2014). A technical annexe is provided alongside the final report, with the detailed methodology for environmental assessment, references and data sources (OECD, 2014).
14. This study uses the IPCC tier-3 approach, with country-specific coefficients from the Switzerland's Greenhouse Gas Inventory 1990-2010 (Bretscher, 2012; OFEN, 2012) and, for dairy cattle, coefficients corrected for milk productivity. For consistency purposes, N_2O emissions from agricultural soils are computed from the nitrogen inputs of the GNB (e.g. inorganic fertilizer, manure, deposition) (See OECD, 2014).
15. For GWP at 100 years, equivalences are 21 tons of CO_2 per ton of CH_4 and 310 for N_2O (UNFCC, 2014).
16. The projected average annual budget for direct payment in 2014 and 2017 is CHF 2 785 million, which is slightly lower than the level in 2012 (CHF 2 804 million). The total amount of payments modelled in the PEM does not match these figures because the model does not represent certain direct payment such as the payments based on non-commodity criteria.
17. The policy impacts on production are simulated assuming all the exogenous factors as constant at 2012 level except for policy change. The production may increase in practice due to higher world price or change in demand structure in Switzerland or overseas.
18. Although the agricultural budget is expected to be constant in 2014-17, the tax payers cost modelled in the OECD PEM is reduced due to the partial shift to the payments based on non-commodity criteria, which are not included in the model.
19. The share of rented land was 43% in plain region, 37% in hilly region and 35% in mountain region in 2012. (Source: Agroscope, Swiss FADN)
20. Producer prices of what, coarse grain and oilseeds presented in Figure 3.15 are calibrated from PSE/CSE database according to PEM commodity group definition (see Table 1 of the technical document *www2.oecd.org/oecdinfo/info.aspx?app=OLIScoteEN&Ref=TAD/CA/APM/WP(2014)28FINAL*. These prices are composite prices of different type of wheat, coarse grain and oilseeds, including both feed and non-feed uses.

21. The simulation equalizes the domestic market price in Switzerland and the EU. However, in practice, the Swiss market price could remain higher for certain commodities due to factors such as quality difference and consumer's preference towards the domestic products.
22. The loss of producer's welfare could be as low as CHF 754 million in case all the rented land is owned by non-farmers.
23. The simulation result shows slightly positive impacts on coarse grain and oilseeds production because relative profitability improvements in coarse grain and oilseeds among arable crop induce land use shift from wheat to these crops. The model assumes area payments for oilseed remains the same level as 2012.
24. This is mainly driven by an increase in domestic consumption due to an approximately 50% lower domestic price. Under the alternative assumption that beef demand is less elastic to price change, the self-sufficiency rate of beef would be 59% and 62% with and without complementary payment, respectively.
25. The simulation assumed the domestic demand for fluid milk is fully satisfied by domestic production. The increase in import of milk in the PEM is in a form of manufactured milk such as SMP, cheese and butter.

References

Aviron, S., H. Nitsch, P. Jeanneret, S. Buholzer, H. Luka, L. Pfiffner, S. Pozzi, B. Schüpbach, T. Walter and F. Herzog (2009), "Ecological cross-compliance promotes farmland biodiversity in Switzerland", *Frontiers in Ecology and Environment*, Vol. 7, pp. 247-252.

Bretscher, D. (2012), "Agricultural CH_4 and N_2O emissions in Switzerland QA/QC", Ettenhausen, Agroscope Reckenholz Tänikon Research Station (ART).

Confédération Suisse (2014), "Ouverture sectorielle réciproque du marché avec l'UE pour tous les produits laitiers", *Rapport du Conseil fédéral*, Confédération Suisse: 112, Berne.

Confédération Suisse (2012), "Évaluation et répercussions de l'accord de libre-échange du fromage entre la Suisse et l'UE", Résumé de l'évaluation effectuée par BAKBASEL sur mandat de l'Office fédéral de l'agriculture, Confédération Suisse, Office Fédéral de l'Agriculture: 6, Berne.

Confédération Suisse (2008), Négociations Suisse-UE pour un accord de libre-échange dans le domaine agro-alimentaire (ALEA); Négociations Suisse-UE pour un accord dans le domaine de la santé publique (ASP). Résultats de l'exploration et analyse. Berne, Confédération Suisse, Département fédéral de l'intérieur DFI, Département fédéral des affaires étrangères DFAE, Département fédéral de l'économie DFE: 45.

Eurostat (2013), "Methodology and Handbook Eurostat/OECD. Nutrient Budgets, EU-27, Norway, Switzerland", Version 1.02, Commission européenne, Eurostat: 112, Luxembourg.

FOEN (2012), "Switzerland's Greenhouse Gas Inventory 1990–2010. National Inventory Report 2012 including reporting elements under the Kyoto Protocol", Submission of 13 April 2012 under the United Nations Framework Convention on Climate Change and under the Kyoto Protocol Bern, Federal Office for the Environment.

Hartmann, M., R. Huber, S. Peter and B. Lehmann (2009), "Strategies to mitigate greenhouse gas and nitrogen emissions in Swiss agriculture: the application of an integrated sector model", *IED Working Paper* 9: 28, ETH Zurich, Institute for Environmental Decisions IED, Zurich.

Herzog, F., V. Prasuhn, E. Spiess and W. Richner (2008), "Environmental cross-compliance mitigates nitrogen and phosphorus pollution from Swiss agriculture", *Environmental Science and Policy*, Vol. 2, pp. 655-668.

IPCC (2006), *IPCC Guidelines for National Greenhouse Gas Inventories*, Intergovernmental Panel on Climate Change.

Kohler, F. (2014b), "Metadata template nutrient budgets », Office fédéral de la statistique, Berne.

Martini, R. (2011), "Long Term Trends in Agricultural Policy Impacts", *OECD Food, Agriculture and Fisheries Papers*, No. 45, OECD Publishing, Paris, *http://dx.doi.org/10.1787/5kgdp5zw179q-en*.

OECD (2014), *Assessing the environmental impacts of agricultural policies with PEM: framework and application to Switzerland* [TAD/CA/APM/WP(2014)28FINAL], Paris, France, OECD.

OECD (2013a), *OECD Compendium of Agri-environmental Indicators*, OECD Publishing, Paris, *http://dx.doi.org/10.1787/9789264186217-en.*

OECD (2013b), *Agricultural Policy Monitoring and Evaluation 2013: OECD Countries and Emerging Economies*, OECD Publishing, Paris, *http://dx.doi.org/10.1787/agr_pol-2013-en.*

OFAG (2013), "Rapport agricole 2013", Office Fédéral de l'Agriculture, Berne.

Peter, S., M. Hartmann, M. Weber, B. Lehmann and W. Hediger (2009), "THG 2020 – Möglichkeiten und Grenzen zur Vermeidung landwirtschaftlicher Treibhausgase in der Schweiz", ETH Zürich, Institut für Umweltentscheidungen: 142, Zürich.

Spiess, E. (2011), "Nitrogen, phosphorus and potassium balances and cycles of Swiss agriculture from 1975 to 2008", *Nutrient Cycling in Agroecosystems*, Vol. 91, pp. 351-365.

UNFCCC (2014), Global Warming Potential referenced to the updated decay response for the Bern carbon cycle model and future CO2 atmospheric concentrations held constant at current levels, *http://unfccc.int/ghg_data/items/3825.php* (accessed on 20 January 2014).

Zimmermann, A., A. Möhring, G. Mack, S. Mann, A. Ferjani and M.-P. Gennaio Franscini (2011), "Les conséquences d'une réforme du système des paiements directs : Simulations à l'aide de modèles SILAS et SWISSland", *Rapport ART 744*, Agroscope Reckenholz-Tänikon ART: 16.

ANNEX 4.A1

Environmental assessment of policy reforms

Table 4.A1.1. Environmental assessment of historical policy reforms: Switzerland

Variable	Region	Unit	RP 93-98 Actual	RP 99-03 Actual	RP 04-07 Actual	RP 08-12 Actual	RP 93-98 No-reform	RP 99-03 No-reform	RP 04-07 No-reform	RP 08-12 No-reform	RP 93-98 Change	RP 99-03 Change	RP 04-07 Change	RP 08-12 Change	RP 93-98 % change	RP 99-03 % change	RP 04-07 % change	RP 08-12 % change
Chemical quantities (main crops)	CHE	1 000 tons a.i.	0.61	0.39	0.38	0.35	0.78	0.71	0.42	0.46	-0.18	-0.32	-0.04	-0.10	-23	-45	-10	-22
N synthetic fertilizer quantities (main crops)	CHE	1 000 tons N	39.61	24.16	20.57	18.08	50.96	44.45	23.48	22.88	-11.35	-20.30	-2.91	-4.80	-22	-46	-12	-21
N synthetic fertilizer quantities (other land)	CHE	1 000 tons N	22.24	24.29	24.73	26.00	21.44	23.87	24.68	25.90	0.80	0.41	0.05	0.11	4	2	0	0
N synthetic fertilizer quantities (total)	CHE	1 000 tons N	61.85	48.44	45.31	44.09	72.40	68.33	48.17	48.78	-10.55	-19.88	-2.86	-4.69	-15	-29	-6	-10
N biological fixation	CHE	1 000 tons N	31.40	32.28	32.62	32.61	31.86	31.89	32.42	32.62	-0.46	0.39	0.20	-0.01	-1	1	1	0
N atmospheric deposition	CHE	1 000 tons N	20.50	20.44	20.31	20.11	20.47	20.42	20.31	20.11	0.03	0.02	0.01	0.00	0	0	0	0
N manure from dairy cattle	CHE	1 000 tons N	75.52	71.84	69.25	68.94	75.58	80.20	67.28	65.18	-0.07	-8.35	1.97	3.76	0	-10	3	6
N manure from beef cattle	CHE	1 000 tons N	34.65	33.47	35.22	38.07	36.86	30.40	31.25	35.25	-2.22	3.07	3.97	2.82	-6	10	13	8
N manure from other livestock	CHE	1 000 tons N	34.57	32.67	33.22	33.38	34.57	32.67	33.22	33.38	0.00	0.00	0.00	0.00	0	0	0	0
N manure (total)	CHE	1 000 tons N	144.73	137.98	137.69	140.38	147.02	143.27	131.75	133.81	-2.29	-5.29	5.94	6.57	-2	-4	5	5
N inputs	CHE	1 000 tons N	258.48	239.15	235.93	237.19	271.75	263.91	232.64	235.31	-13.27	-24.76	3.29	1.87	-5	-9	1	1
N uptakes (main crops)	CHE	1 000 tons N	21.17	17.40	17.99	17.13	27.98	33.05	21.47	21.59	-6.80	-15.64	-3.48	-4.45	-24	-47	-16	-21
N uptakes (other land)	CHE	1 000 tons N	132.41	132.71	132.99	133.44	132.56	131.00	132.35	133.26	-0.14	1.70	0.63	0.17	0	1	0	0
N uptakes (total)	CHE	1 000 tons N	153.59	150.11	150.98	150.57	160.53	164.05	153.83	154.85	-6.95	-13.94	-2.85	-4.28	-4	-8	-2	-3
Gross Nitrogen Balance	CHE	1 000 tons N	104.89	89.04	84.95	86.62	111.21	99.86	78.82	80.46	-6.32	-10.82	6.13	6.16	-6	-11	8	8
P synthetic fertilizer quantities (main crops)	CHE	1 000 tons P	6.61	4.03	3.43	3.02	8.51	7.42	3.92	3.82	-1.89	-3.39	-0.49	-0.80	-22	-46	-12	-21
P synthetic fertilizer quantities (other land)	CHE	1 000 tons P	3.54	3.60	3.61	3.75	3.23	3.52	3.63	3.72	0.32	0.08	-0.01	0.04	10	2	0	1
P synthetic fertilizer quantities (total)	CHE	1 000 tons P	10.16	7.63	7.04	6.77	11.73	10.94	7.54	7.54	-1.58	-3.31	-0.50	-0.77	-13	-30	-7	-10
P atmospheric deposition	CHE	1 000 tons P	0.39	0.31	0.30	0.29	0.39	0.31	0.30	0.29	0.00	0.00	0.00	0.00	0	0	0	0
P manure from dairy cattle	CHE	1 000 tons P	11.93	10.69	10.34	10.23	11.95	11.85	10.05	9.62	-0.02	-1.16	0.29	0.61	0	-10	3	6
P manure from beef cattle	CHE	1 000 tons P	5.03	4.93	5.19	5.57	5.35	4.48	4.61	5.16	-0.32	0.45	0.59	0.41	-6	10	13	8
P manure from other livestock	CHE	1 000 tons P	9.33	9.28	9.78	9.78	9.33	9.28	9.78	9.78	0.00	0.00	0.00	0.00	0	0	0	0

Table 4.A1.1. **Environmental assessment of historical policy reforms: Switzerland** *(cont.)*

Variable	Region	Unit	RP 93-98 Actual	RP 99-03 Actual	RP 04-07 Actual	RP 08-12 Actual	RP 93-98 No-reform	RP 99-03 No-reform	RP 04-07 No-reform	RP 08-12 No-reform	RP 93-98 Change	RP 99-03 Change	RP 04-07 Change	RP 08-12 Change	RP 93-98 % change	RP 99-03 % change	RP 04-07 % change	RP 08-12 % change
P manure (total)	CHE	1 000 tons P	26.29	24.90	25.31	25.58	26.63	25.61	24.43	24.56	-0.34	-0.71	0.88	1.02	-1	-3	4	4
P inputs	CHE	1 000 tons P	36.83	32.85	32.65	32.64	38.75	36.87	32.27	32.39	-1.92	-4.02	0.38	0.26	-5	-11	1	1
P uptakes (main crops)	CHE	1 000 tons P	4.19	3.42	3.56	3.39	5.64	6.42	4.23	4.33	-1.45	-3.00	-0.66	-0.95	-26	-47	-16	-22
P uptakes (other land)	CHE	1 000 tons P	21.27	21.84	21.99	22.26	21.26	21.57	21.89	22.23	0.01	0.27	0.10	0.03	0	1	0	0
P uptakes (total)	CHE	1 000 tons P	25.46	25.26	25.55	25.64	26.89	27.99	26.12	26.56	-1.44	-2.73	-0.57	-0.92	-5	-10	-2	-3
Gross Phosphorous Balance	CHE	1 000 tons P	11.38	7.59	7.10	7.00	11.85	8.88	6.16	5.83	-0.48	-1.29	0.94	1.17	-4	-15	15	20
CH_4 enteric fermentation (dairy)	CHE	1 000 tons CH_4	85.82	80.94	79.30	81.36	85.90	90.26	77.10	77.12	-0.08	-9.32	2.19	4.25	0	-10	3	6
CH_4 management (dairy)	CHE	1 000 tons CH_4	16.29	15.39	15.41	15.96	16.31	17.16	14.99	15.12	-0.02	-1.77	0.42	0.83	0	-10	3	6
CH_4 emissions (dairy)	CHE	1 000 tons CH_4	102.11	96.32	94.71	97.32	102.21	107.42	92.09	92.24	-0.10	-11.10	2.61	5.08	0	-10	3	6
CH_4 enteric fermentation (viande)	CHE	1 000 tons CH_4	35.53	34.63	36.00	38.78	37.77	31.45	31.94	35.91	-2.24	3.18	4.06	2.87	-6	10	13	8
CH_4 manure management (beef)	CHE	1 000 tons CH_4	4.40	4.20	4.54	5.02	4.67	3.82	4.03	4.65	-0.27	0.38	0.51	0.37	-6	10	13	8
CH_4 emissions (beef)	CHE	1 000 tons CH_4	39.93	38.83	40.53	43.80	42.44	35.26	35.96	40.55	-2.51	3.57	4.57	3.24	-6	10	13	8
CH_4 emissions (cattle)	CHE	1 000 tons CH_4	142.04	135.16	135.24	141.12	144.65	142.68	128.06	132.79	-2.61	-7.53	7.19	8.32	-2	-5	6	6
CH_4 other livestock	CHE	1 000 tons CH_4	18.31	18.52	19.43	19.31	18.31	18.52	19.43	19.31	0.00	0.00	0.00	0.00	0	0	0	0
CH_4 emissions (total)	CHE	1 000 tons CH_4	160.35	153.68	154.67	160.42	162.96	161.20	147.48	152.10	-2.61	-7.53	7.19	8.32	-2	-5	5	5
N_2O manure management	CHE	1 000 tons N_2O	1.34	1.17	1.09	1.11	1.36	1.19	1.05	1.06	-0.03	-0.02	0.05	0.05	-2	-2	5	4
N_2O direct soil emissions	CHE	1 000 tons N_2O	3.95	3.57	3.52	3.54	4.22	4.06	3.52	3.59	-0.28	-0.49	0.00	-0.05	-7	-12	0	-2
N_2O pasture range and paddock	CHE	1 000 tons N_2O	0.51	0.76	0.82	0.84	0.52	0.79	0.79	0.81	-0.01	-0.03	0.04	0.04	-2	-4	5	5
N_2O athmospheric deposition	CHE	1 000 tons N_2O	0.32	0.32	0.32	0.32	0.32	0.32	0.32	0.32	0.00	0.00	0.00	0.00	0	0	0	0
N_2O leaching and run-off	CHE	1 000 tons N_2O	1.62	1.46	1.44	1.45	1.72	1.66	1.41	1.43	-0.10	-0.20	0.02	0.01	-6	-12	2	1
N_2O emissions (total)	CHE	1 000 tons N_2O	7.74	7.28	7.19	7.25	8.16	8.02	7.09	7.21	-0.41	-0.74	0.11	0.05	-5	-9	2	1
Global warming potential	CHE	1 000 tons eqCO_2	5 768.22	5 483.65	5 477.70	5 617.13	5 950.36	5 870.77	5 293.80	5 428.30	-182.14	-387.12	183.90	188.83	-3	-7	3	3

Source: Model simulation with OECD Policy Evaluation Model.

Table 4.A1.2. Environmental assessment of AP 14-17 and further reforms: Switzerland

Variable	Region	Unit	2012 Actual	AP 14-17 Reform	EU market integration Reform	With compensation Reform	AP 14-17 Change	EU market integration Change	With compensation Change	AP 14-17 % change	EU market integration % change	With compensation % change
Chemical quantities (main crops)	CHE	1 000 tons a.i.	0.34	0.34	0.34	0.34	0.00	0.00	0.00	0	1	1
N synthetic fertilizer quantities (main crops)	CHE	1 000 tons N	17.76	17.86	17.60	17.80	0.09	-0.16	0.04	1	-1	0
N synthetic fertilizer quantities (other land)	CHE	1 000 tons N	26.23	26.31	27.12	26.86	0.08	0.89	0.63	0	3	2
N synthetic fertilizer quantities (total)	CHE	1 000 tons N	43.99	44.17	44.71	44.66	0.17	0.72	0.67	0	2	2
N biological fixation	CHE	1 000 tons N	32.54	32.35	31.81	32.01	-0.19	-0.72	-0.53	-1	-2	-2
N atmospheric deposition	CHE	1 000 tons N	20.06	20.05	19.92	20.00	-0.01	-0.15	-0.06	0	-1	0
N manure from dairy cattle	CHE	1 000 tons N	68.44	64.55	63.89	64.05	-3.89	-4.55	-4.40	-6	-7	-6
N manure from beef cattle	CHE	1 000 tons N	37.69	30.85	34.01	35.58	-6.85	-3.68	-2.11	-18	-10	-6
N manure from other livestock	CHE	1 000 tons N	33.26	33.26	33.26	33.26	0.00	0.00	0.00	0	0	0
N manure (total)	CHE	1 000 tons N	139.39	128.65	131.16	132.88	-10.74	-8.24	-6.51	-8	-6	-5
N inputs	CHE	1 000 tons N	235.98	225.22	227.60	229.56	-10.76	-8.38	-6.43	-5	-4	-3
N uptakes (main crops)	CHE	1 000 tons N	16.55	16.69	16.15	16.41	0.14	-0.40	-0.13	1	-2	-1
N uptakes (other land)	CHE	1 000 tons N	132.85	132.44	131.95	132.24	-0.41	-0.90	-0.60	0	-1	0
N uptakes (total)	CHE	1 000 tons N	149.39	149.13	148.10	148.66	-0.27	-1.29	-0.74	0	-1	0
Gross Nitrogen Balance	CHE	1 000 tons N	86.59	76.10	79.50	80.90	-10.49	-7.09	-5.69	-12	-8	-7
P synthetic fertilizer quantities (main crops)	CHE	1 000 tons P	2.96	2.98	2.94	2.97	0.02	-0.03	0.01	1	-1	0
P synthetic fertilizer quantities (other land)	CHE	1 000 tons P	3.75	3.80	4.12	4.03	0.05	0.37	0.28	1	10	7
P synthetic fertilizer quantities (total)	CHE	1 000 tons P	6.71	6.78	7.05	7.00	0.07	0.34	0.29	1	5	4
P atmospheric deposition	CHE	1 000 tons P	0.29	0.29	0.29	0.29	0.00	0.00	0.00	0	-1	0
P manure from dairy cattle	CHE	1 000 tons P	10.21	9.53	9.58	9.60	-0.68	-0.64	-0.61	-7	-6	-6
P manure from beef cattle	CHE	1 000 tons P	5.51	4.51	4.98	5.20	-1.00	-0.54	-0.31	-18	-10	-6
P manure from other livestock	CHE	1 000 tons P	9.65	9.65	9.65	9.65	0.00	0.00	0.00	0	0	0

Table 4.A1.2. **Environmental assessment of AP 14-17 and further reforms: Switzerland** *(cont.)*

Variable	Region	Unit	2012 Actual	AP 14-17 Reform	EU market integration Reform	With compensation Reform	AP 14-17 Change	EU market integration Change	With compensation Change	AP 14-17 % change	EU market integration % change	With compensation % change
P manure (total)	CHE	1 000 tons P	25.37	23.69	24.20	24.45	-1.68	-1.18	-0.92	-7	-5	-4
P inputs	CHE	1 000 tons P	32.38	30.76	31.54	31.74	-1.61	-0.84	-0.64	-5	-3	-2
P uptakes (main crops)	CHE	1 000 tons P	3.27	3.30	3.23	3.27	0.03	-0.04	0.00	1	-1	0
P uptakes (other land)	CHE	1 000 tons P	22.30	22.23	22.15	22.21	-0.06	-0.15	-0.09	0	-1	0
P uptakes (total)	CHE	1 000 tons P	25.57	25.53	25.38	25.48	-0.04	-0.18	-0.09	0	-1	0
Gross Phosphorous Balance	CHE	1 000 tons P	6.81	5.23	6.16	6.26	-1.57	-0.65	-0.55	-23	-10	-8
CH_4 enteric fermentation (dairy)	CHE	1 000 tons CH_4	81.28	77.08	75.72	75.94	-4.21	-5.56	-5.34	-5	-7	-7
CH_4 management (dairy)	CHE	1 000 tons CH_4	15.94	15.11	14.85	14.89	-0.82	-1.09	-1.05	-5	-7	-7
CH_4 emissions (dairy)	CHE	1 000 tons CH_4	97.22	92.19	90.56	90.83	-5.03	-6.66	-6.38	-5	-7	-7
CH_4 enteric fermentation (viande)	CHE	1 000 tons CH_4	38.53	31.53	34.77	36.37	-6.99	-3.76	-2.15	-18	-10	-6
CH_4 manure management (beef)	CHE	1 000 tons CH_4	5.01	4.10	4.52	4.73	-0.91	-0.49	-0.28	-18	-10	-6
CH_4 emissions (beef)	CHE	1 000 tons CH_4	43.54	35.63	39.29	41.10	-7.90	-4.25	-2.43	-18	-10	-6
CH_4 emissions (cattle)	CHE	1 000 tons CH_4	140.76	127.82	129.85	131.94	-12.94	-10.90	-8.82	-9	-8	-6
CH_4 other livestock	CHE	1 000 tons CH_4	19.17	19.17	19.17	19.17	0.00	0.00	0.00	0	0	0
CH_4 emissions (total)	CHE	1 000 tons CH_4	159.93	146.99	149.03	151.11	-12.94	-10.90	-8.82	-8	-7	-6
N_2O manure management	CHE	1 000 tons N_2O	1.11	1.02	1.05	1.06	-0.08	-0.06	-0.04	-7	-5	-4
N_2O direct soil emissions	CHE	1 000 tons N_2O	3.53	2.98	3.01	3.03	-0.54	-0.52	-0.50	-15	-15	-14
N_2O pasture range and paddock	CHE	1 000 tons N_2O	0.84	0.78	0.79	0.80	-0.06	-0.05	-0.04	-8	-6	-5
N_2O athmospheric deposition	CHE	1 000 tons N_2O	0.32	0.32	0.31	0.31	0.00	0.00	0.00	0	-1	0
N_2O leaching and run-off	CHE	1 000 tons N_2O	1.44	1.36	1.38	1.39	-0.08	-0.06	-0.05	-6	-4	-3
N_2O emissions (total)	CHE	1 000 tons N_2O	7.23	6.46	6.54	6.60	-0.77	-0.69	-0.63	-11	-10	-9
Global warming potential	CHE	1 000 tons $eqCO_2$	5 599.54	5 222.15	5 293.04	5 355.80	-377.39	-306.51	-243.74	-7	-5	-4

Source: Model simulation with OECD Policy Evaluation Model.

Table 4.A1.3. Environmental assessment of historical policy reforms: Plain region

Variable	Region	Unit	RP 93-98 Actual	RP 99-03 Actual	RP 04-07 Actual	RP 08-12 Actual	RP 93-98 No-reform	RP 99-03 No-reform	RP 04-07 No-reform	RP 08-12 No-reform	RP 93-98 Change	RP 99-03 Change	RP 04-07 Change	RP 08-12 Change	RP 93-98 % change	RP 99-03 % change	RP 04-07 % change	RP 08-12 % change
Chemical quantities (main crops)	chp	1000 tons a.i.	0.51	0.33	0.33	0.31	0.67	0.60	0.36	0.40	-0.15	-0.27	-0.04	-0.09	-23	-45	-10	-23
N synthetic fertilizer quantities (main crops)	chp	1000 tons N	33.22	20.37	17.59	15.55	42.96	37.49	20.06	19.71	-9.74	-17.12	-2.47	-4.16	-23	-46	-12	-21
N synthetic fertilizer quantities (other land)	chp	1000 tons N	15.64	16.78	17.04	17.99	15.06	16.36	16.97	17.89	0.58	0.42	0.06	0.10	4	3	0	1
N synthetic fertilizer quantities (total)	chp	1000 tons N	48.86	37.15	34.63	33.54	58.03	53.85	37.03	37.60	-9.16	-16.71	-2.40	-4.06	-16	-31	-6	-11
N biological fixation	chp	1000 tons N	15.73	16.03	16.25	16.29	16.02	15.70	16.08	16.30	-0.29	0.33	0.17	-0.01	-2	2	1	0
N atmospheric deposition	chp	1000 tons N	10.23	10.12	10.01	9.90	10.21	10.10	10.01	9.90	0.02	0.02	0.00	0.00	0	0	0	0
N manure from dairy cattle	chp	1000 tons N	35.24	33.21	31.29	31.05	35.03	37.00	30.70	30.19	0.21	-3.79	0.59	0.86	1	-10	2	3
N manure from beef cattle	chp	1000 tons N	15.20	13.93	14.30	15.67	16.57	13.26	12.69	15.57	-1.37	0.67	1.61	0.10	-8	5	13	1
N manure from other livestock	chp	1000 tons N	19.87	17.95	18.40	18.67	19.87	17.95	18.40	18.67	0.00	0.00	0.00	0.00	0	0	0	0
N manure (total)	chp	1000 tons N	70.30	65.10	63.98	65.39	71.47	68.21	61.79	64.43	-1.17	-3.12	2.19	0.96	-2	-5	4	1
N inputs	chp	1000 tons N	145.13	128.39	124.88	125.13	155.73	147.86	124.91	128.23	-10.60	-19.47	-0.04	-3.10	-7	-13	0	-2
N uptakes (main crops)	chp	1000 tons N	17.60	14.58	15.30	14.67	23.37	27.76	18.27	18.48	-5.77	-13.19	-2.97	-3.81	-25	-47	-16	-21
N uptakes (other land)	chp	1000 tons N	67.86	67.30	67.40	68.29	67.69	65.77	66.85	68.13	0.17	1.53	0.55	0.16	0	2	1	0
N uptakes (total)	chp	1000 tons N	85.45	81.87	82.70	82.96	91.06	93.53	85.12	86.61	-5.60	-11.66	-2.42	-3.65	-6	-12	-3	-4
Gross Nitrogen Balance	chp	1000 tons N	59.68	46.52	42.17	42.17	64.67	54.33	39.80	41.62	-4.99	-7.81	2.38	0.55	-8	-14	6	1
Gross Nitrogen Balance per hectare	chp	Kg N/ha	116.11	91.46	83.82	84.75	125.81	106.82	79.10	83.65	-9.70	-15.36	4.73	1.10	-8	-14	6	1
P synthetic fertilizer quantities (main crops)	chp	1000 tons P	5.55	3.40	2.94	2.60	7.17	6.26	3.35	3.29	-1.63	-2.86	-0.41	-0.69	-23	-46	-12	-21
P synthetic fertilizer quantities (other land)	chp	1000 tons P	2.78	2.87	2.90	3.03	2.55	2.79	2.91	3.00	0.23	0.08	-0.01	0.03	9	3	0	1
P synthetic fertilizer quantities (total)	chp	1000 tons P	8.33	6.27	5.84	5.63	9.72	9.05	6.26	6.29	-1.39	-2.77	-0.42	-0.66	-14	-31	-7	-11
P atmospheric deposition	chp	1000 tons P	0.18	0.15	0.14	0.14	0.18	0.15	0.14	0.14	0.00	0.00	0.00	0.00	0	0	0	0
P manure from dairy cattle	chp	1000 tons P	5.33	4.69	4.50	4.50	5.31	5.23	4.41	4.38	0.02	-0.54	0.08	0.13	0	-10	2	3
P manure from beef cattle	chp	1000 tons P	2.21	2.05	2.11	2.29	2.41	1.96	1.87	2.28	-0.20	0.10	0.24	0.01	-8	5	13	1
P manure from other livestock	chp	1000 tons P	4.72	4.55	4.89	4.98	4.72	4.55	4.89	4.98	0.00	0.00	0.00	0.00	0	0	0	0

Table 4.A1.3. Environmental assessment of historical policy reforms: Plain region (*cont.*)

Variable	Region	Unit	RP 93-98 Actual	RP 99-03 Actual	RP 04-07 Actual	RP 08-12 Actual	RP 93-98 No-reform	RP 99-03 No-reform	RP 04-07 No-reform	RP 08-12 No-reform	RP 93-98 Change	RP 99-03 Change	RP 04-07 Change	RP 08-12 Change	RP 93-98 % change	RP 99-03 % change	RP 04-07 % change	RP 08-12 % change
P manure (total)	chp	1000 tons P	12.25	11.30	11.49	11.78	12.43	11.74	11.18	11.64	-0.18	-0.44	0.32	0.14	-1	-4	3	1
P inputs	chp	1000 tons P	20.76	17.72	17.47	17.54	22.33	20.93	17.57	18.06	-1.57	-3.21	-0.10	-0.52	-7	-15	-1	-3
P uptakes (main crops)	chp	1000 tons P	3.46	2.85	3.02	2.90	4.69	5.37	3.59	3.71	-1.23	-2.52	-0.56	-0.81	-26	-47	-16	-22
P uptakes (other land)	chp	1000 tons P	10.23	10.52	10.60	10.85	10.17	10.28	10.51	10.82	0.06	0.24	0.08	0.03	1	2	1	0
P uptakes (total)	chp	1000 tons P	13.69	13.37	13.62	13.75	14.86	15.65	14.10	14.53	-1.17	-2.28	-0.48	-0.78	-8	-15	-3	-5
Gross Phosphorous Balance	chp	1000 tons P	7.07	4.35	3.85	3.79	7.48	5.29	3.47	3.53	-0.40	-0.94	0.38	0.26	-5	-18	11	7
Gross Phosphorous Balance per hectare	chp	Kg P/ha	13.76	8.55	7.65	7.62	14.55	10.39	6.89	7.09	-0.79	-1.85	0.76	0.53	-5	-18	11	7
CH_4 enteric fermentation (dairy)	chp	1000 tons CH_4	39.85	37.32	36.32	37.51	39.62	41.55	35.63	36.49	0.23	-4.23	0.69	1.02	1	-10	2	3
CH_4 management (dairy)	chp	1000 tons CH_4	7.56	7.09	7.06	7.36	7.52	7.90	6.93	7.16	0.04	-0.80	0.13	0.20	1	-10	2	3
CH_4 emissions (dairy)	chp	1000 tons CH_4	47.41	44.41	43.38	44.87	47.14	49.45	42.56	43.65	0.27	-5.04	0.82	1.22	1	-10	2	3
CH_4 enteric fermentation (beef)	chp	1000 tons CH_4	15.59	14.35	14.61	15.95	16.98	13.66	12.97	15.85	-1.39	0.69	1.64	0.10	-8	5	13	1
CH_4 management (beef)	chp	1000 tons CH_4	1.93	1.74	1.84	2.06	2.10	1.66	1.63	2.05	-0.17	0.08	0.21	0.01	-8	5	13	1
CH_4 emissions (beef)	chp	1000 tons CH_4	17.52	16.09	16.45	18.02	19.08	15.31	14.61	17.90	-1.56	0.78	1.85	0.12	-8	5	13	1
CH_4 emissions (cattle)	chp	1000 tons CH_4	64.93	60.50	59.84	62.89	66.22	64.76	57.16	61.55	-1.29	-4.26	2.67	1.34	-2	-7	5	2
CH_4 other livestock	chp	1000 tons CH_4	9.48	9.54	10.19	10.21	9.48	9.54	10.19	10.21	0.00	0.00	0.00	0.00	0	0	0	0
CH_4 emissions (total)	chp	1000 tons CH_4	74.41	70.03	70.03	73.10	75.70	74.29	67.36	71.76	-1.29	-4.26	2.67	1.34	-2	-6	4	2
N_2O manure management	chp	1000 tons N_2O	0.62	0.52	0.49	0.51	0.63	0.54	0.47	0.50	-0.01	-0.02	0.02	0.01	-2	-3	4	1
N_2O direct soil emissions	chp	1000 tons N_2O	2.27	1.97	1.92	1.93	2.50	2.36	1.95	2.02	-0.23	-0.40	-0.03	-0.10	-9	-17	-2	-5
N_2O pasture range and paddock	chp	1000 tons N_2O	0.25	0.36	0.38	0.39	0.25	0.37	0.37	0.39	0.00	-0.02	0.01	0.01	-2	-5	4	1
N_2O athmospheric deposition	chp	1000 tons N_2O	0.16	0.16	0.16	0.16	0.16	0.16	0.16	0.16	0.00	0.00	0.00	0.00	0	0	0	0
N_2O leaching and run-off	chp	1000 tons N_2O	0.94	0.80	0.77	0.78	1.02	0.96	0.78	0.80	-0.08	-0.16	0.00	-0.02	-8	-16	0	-3
N_2O emissions (total)	chp	1000 tons N_2O	4.24	3.81	3.73	3.76	4.56	4.40	3.73	3.87	-0.33	-0.59	0.00	-0.11	-7	-13	0	-3
Global warming potential	chp	1000 tons $eqCO_2$	2876.45	2651.99	2626.63	2700.49	3004.43	2923.16	2570.50	2706.03	-127.98	-271.17	56.13	-5.54	-4	-9	2	0

Source: Model simulation with OECD Policy Evaluation Model.

Table 4.A1.4. Environmental assessment of AP 14-17 and further reforms: Plain region

Variable	Region	Unit	2012 Actual	AP 14-17 Reform	EU market integration Reform	With compensation Reform	AP 14-17 Change	EU market integration Change	With compensation Change	AP 14-17 % change	EU market integration % change	With compensation % change
Chemical quantities (main crops)	chp	1 000 tons a.i.	0.29	0.29	0.28	0.29	0.00	0.01	0.00	0	-3	-1
N synthetic fertilizer quantities (main crops)	chp	1 000 tons N	15.26	15.33	14.49	14.85	-0.07	0.76	0.41	0	-5	-3
N synthetic fertilizer quantities (other land)	chp	1 000 tons N	18.17	18.21	18.68	18.56	-0.04	-0.52	-0.39	0	3	2
N synthetic fertilizer quantities (total)	chp	1 000 tons N	33.42	33.54	33.18	33.40	-0.12	0.24	0.02	0	-1	0
N biological fixation	chp	1 000 tons N	16.28	16.12	16.16	16.16	0.16	0.12	0.12	-1	-1	-1
N atmospheric deposition	chp	1 000 tons N	9.86	9.86	9.80	9.84	0.00	0.06	0.03	0	-1	0
N manure from dairy cattle	chp	1 000 tons N	30.73	29.80	28.34	28.27	0.93	2.38	2.46	-3	-8	-8
N manure from beef cattle	chp	1 000 tons N	15.54	13.55	12.84	13.29	1.99	2.70	2.25	-13	-17	-14
N manure from other livestock	chp	1 000 tons N	18.77	18.77	18.77	18.77	0.00	0.00	0.00	0	0	0
N manure (total)	chp	1 000 tons N	65.03	62.11	59.95	60.32	2.92	5.08	4.71	-4	-8	-7
N inputs	chp	1 000 tons N	124.60	121.64	119.09	119.73	2.96	5.51	4.87	-2	-4	-4
N uptakes (main crops)	chp	1 000 tons N	14.16	14.28	13.20	13.60	-0.12	0.96	0.56	1	-7	-4
N uptakes (other land)	chp	1 000 tons N	68.12	67.80	68.66	68.44	0.32	-0.54	-0.33	0	1	0
N uptakes (total)	chp	1 000 tons N	82.28	82.08	81.85	82.05	0.20	0.42	0.23	0	-1	0
Gross Nitrogen Balance	chp	1 000 tons N	42.33	39.56	37.24	37.68	2.76	5.09	4.64	-7	-12	-11
Gross Nitrogen Balance per hectare	chp	Kg N/ha	85.40	79.83	75.13	76.03	5.57	10.27	9.37	-7	-12	-11
P synthetic fertilizer quantities (main crops)	chp	1 000 tons P	2.55	2.56	2.42	2.48	-0.01	0.13	0.07	0	-5	-3
P synthetic fertilizer quantities (other land)	chp	1 000 tons P	3.03	3.07	3.22	3.18	-0.04	-0.19	-0.15	1	6	5
P synthetic fertilizer quantities (total)	chp	1 000 tons P	5.58	5.63	5.64	5.66	-0.06	-0.06	-0.08	1	1	1
P atmospheric deposition	chp	1 000 tons P	0.14	0.14	0.14	0.14	0.00	0.00	0.00	0	-1	0
P manure from dairy cattle	chp	1 000 tons P	4.50	4.35	4.15	4.14	0.14	0.34	0.36	-3	-8	-8
P manure from beef cattle	chp	1 000 tons P	2.27	1.98	1.88	1.94	0.29	0.39	0.33	-13	-17	-14
P manure from other livestock	chp	1 000 tons P	4.97	4.97	4.97	4.97	0.00	0.00	0.00	0	0	0

Table 4.A1.4. **Environmental assessment of AP 14-17 and further reforms: Plain region** *(cont.)*

Variable	Region	Unit	2012 Actual	AP 14-17 Reform	EU market integration Reform	With compensation Reform	AP 14-17 Change	EU market integration Change	With compensation Change	AP 14-17 % change	EU market integration % change	With compensation % change
P manure (total)	chp	1 000 tons P	11.74	11.30	11.00	11.05	0.43	0.74	0.68	-4	-6	-6
P inputs	chp	1 000 tons P	17.45	17.07	16.77	16.85	0.38	0.68	0.60	-2	-4	-3
P uptakes (main crops)	chp	1 000 tons P	2.79	2.82	2.64	2.71	-0.02	0.15	0.08	1	-5	-3
P uptakes (other land)	chp	1 000 tons P	10.90	10.85	10.99	10.96	0.05	-0.10	-0.06	0	1	1
P uptakes (total)	chp	1 000 tons P	13.69	13.67	13.64	13.67	0.02	0.05	0.02	0	0	0
Gross Phosphorous Balance	chp	1 000 tons P	3.76	3.41	3.13	3.18	0.35	0.62	0.58	-9	-17	-15
Gross Phosphorous Balance per hectare	chp	Kg P/ha	7.58	6.87	6.32	6.41	0.71	1.26	1.17	-9	-17	-15
CH_4 enteric fermentation (dairy)	chp	1 000 tons CH_4	37.43	36.45	34.42	34.32	0.98	3.01	3.11	-3	-8	-8
CH_4 management (dairy)	chp	1 000 tons CH_4	7.34	7.15	6.75	6.73	0.19	0.59	0.61	-3	-8	-8
CH_4 emissions (dairy)	chp	1 000 tons CH_4	44.77	43.60	41.17	41.05	1.17	3.60	3.72	-3	-8	-8
CH_4 enteric fermentation (viande)	chp	1 000 tons CH_4	15.85	13.82	13.10	13.55	2.03	2.75	2.30	-13	-17	-14
CH_4 manure management (beef)	chp	1 000 tons CH_4	2.06	1.79	1.70	1.76	0.26	0.36	0.30	-13	-17	-14
CH_4 emissions (beef)	chp	1 000 tons CH_4	17.91	15.62	14.80	15.32	2.29	3.11	2.59	-13	-17	-14
CH_4 emissions (cattle)	chp	1 000 tons CH_4	62.68	59.22	55.97	56.37	3.47	6.71	6.31	-6	-11	-10
CH_4 other livestock	chp	1 000 tons CH_4	10.18	10.18	10.18	10.18	0.00	0.00	0.00	0	0	0
CH_4 emissions (total)	chp	1 000 tons CH_4	72.87	69.40	66.15	66.55	3.47	6.71	6.31	-5	-9	-9
N_2O manure management	chp	1 000 tons N_2O	0.51	0.49	0.47	0.47	0.02	0.04	0.03	-5	-7	-7
N_2O direct soil emissions	chp	1 000 tons N_2O	1.93	1.89	1.87	1.88	0.03	0.05	0.05	-2	-3	-2
N_2O pasture range and paddock	chp	1 000 tons N_2O	0.39	0.37	0.36	0.36	0.02	0.03	0.03	-4	-8	-7
N_2O athmospheric deposition	chp	1 000 tons N_2O	0.15	0.15	0.15	0.15	0.00	0.00	0.00	0	-1	0
N_2O leaching and run-off	chp	1 000 tons N_2O	0.77	0.75	0.73	0.74	0.02	0.04	0.04	-3	-5	-5
N_2O emissions (total)	chp	1 000 tons N_2O	3.75	3.66	3.59	3.61	0.09	0.16	0.15	-2	-4	-4
Global warming potential	chp	1 000 tons eqCO_2	2 694.04	2 592.36	2 502.01	2 516.10	101.68	192.03	177.93	-4	-7	-7

Source: Model simulation with OECD Policy Evaluation Model.

Table 4.A1.5. Environmental assessment of historical policy reforms: Hilly region

Variable	Region	Unit	RP 93-98 Actual	RP 99-03 Actual	RP 04-07 Actual	RP 08-12 Actual	RP 93-98 No-reform	RP 99-03 No-reform	RP 04-07 No-reform	RP 08-12 No-reform	RP 93-98 Change	RP 99-03 Change	RP 04-07 Change	RP 08-12 Change	RP 93-98 % change	RP 99-03 % change	RP 04-07 % change	RP 08-12 % change
Chemical quantities (main crops)	chh	1 000 tons a.i.	0.09	0.05	0.05	0.04	0.11	0.10	0.05	0.05	-0.02	-0.05	-0.01	-0.01	-20	-45	-13	-20
N synthetic fertilizer quantities (main crops)	chh	1 000 tons N	5.88	3.49	2.74	2.34	7.35	6.44	3.19	2.90	-1.47	-2.95	-0.45	-0.57	-20	-46	-14	-20
N synthetic fertilizer quantities (other land)	chh	1 000 tons N	5.55	6.20	6.39	6.70	5.36	6.20	6.40	6.69	0.19	0.00	-0.01	0.01	4	0	0	0
N synthetic fertilizer quantities (total)	chh	1 000 tons N	11.43	9.69	9.13	9.03	12.71	12.64	9.59	9.59	-1.27	-2.95	-0.46	-0.56	-10	-23	-5	-6
N biological fixation	chh	1 000 tons N	9.60	9.98	10.11	10.17	9.75	9.93	10.08	10.17	-0.15	0.05	0.03	0.00	-2	0	0	0
N atmospheric deposition	chh	1 000 tons N	5.13	5.08	5.05	5.00	5.12	5.07	5.05	5.00	0.01	0.00	0.00	0.00	0	0	0	0
N manure from dairy cattle	chh	1 000 tons N	23.93	23.03	22.54	22.10	24.10	25.90	21.75	21.13	-0.18	-2.88	0.79	0.98	-1	-11	4	5
N manure from beef cattle	chh	1 000 tons N	10.02	9.89	10.59	11.44	10.62	8.66	9.11	10.66	-0.60	1.23	1.48	0.78	-6	14	16	7
N manure from other livestock	chh	1 000 tons N	9.57	8.88	9.08	9.11	9.57	8.88	9.08	9.11	0.00	0.00	0.00	0.00	0	0	0	0
N manure (total)	chh	1 000 tons N	43.52	41.79	42.22	42.66	44.30	43.44	39.95	40.90	-0.78	-1.65	2.27	1.76	-2	-4	6	4
N inputs	chh	1 000 tons N	69.68	66.53	66.51	66.86	71.87	71.08	64.67	65.67	-2.19	-4.55	1.84	1.20	-3	-6	3	2
N uptakes (main crops)	chh	1 000 tons N	3.27	2.61	2.51	2.31	4.21	4.90	3.01	2.89	-0.94	-2.29	-0.50	-0.58	-22	-47	-17	-20
N uptakes (other land)	chh	1 000 tons N	39.34	39.64	39.84	39.83	39.57	39.48	39.77	39.82	-0.23	0.16	0.08	0.01	-1	0	0	0
N uptakes (total)	chh	1 000 tons N	42.62	42.25	42.35	42.14	43.78	44.38	42.78	42.71	-1.17	-2.13	-0.43	-0.58	-3	-5	-1	-1
Gross Nitrogen Balance	chh	1 000 tons N	27.07	24.28	24.16	24.72	28.09	26.70	21.89	22.95	-1.02	-2.42	2.27	1.77	-4	-9	10	8
Gross Nitrogen Balance per hectare	chh	Kg N/ha	96.67	87.55	87.55	90.47	100.30	96.27	79.34	83.99	-3.63	-8.72	8.21	6.48	-4	-9	10	8
P synthetic fertilizer quantities (main crops)	chh	1 000 tons P	0.98	0.58	0.46	0.39	1.23	1.07	0.53	0.48	-0.24	-0.49	-0.07	-0.09	-20	-46	-14	-20
P synthetic fertilizer quantities (other land)	chh	1 000 tons P	0.65	0.62	0.62	0.63	0.57	0.63	0.62	0.62	0.08	-0.01	-0.01	0.00	13	-1	-1	1
P synthetic fertilizer quantities (total)	chh	1 000 tons P	1.63	1.21	1.07	1.02	1.80	1.71	1.16	1.11	-0.17	-0.50	-0.08	-0.09	-9	-29	-7	-8
P atmospheric deposition	chh	1 000 tons P	0.10	0.08	0.08	0.08	0.10	0.08	0.08	0.08	0.00	0.00	0.00	0.00	0	0	0	0
P manure from dairy cattle	chh	1 000 tons P	3.69	3.35	3.26	3.23	3.72	3.73	3.15	3.08	-0.02	-0.38	0.12	0.15	-1	-10	4	5
P manure from beef cattle	chh	1 000 tons P	1.45	1.46	1.56	1.67	1.54	1.28	1.34	1.56	-0.09	0.18	0.22	0.11	-6	14	16	7
P manure from other livestock	chh	1 000 tons P	2.50	2.46	2.64	2.63	2.50	2.46	2.64	2.63	0.00	0.00	0.00	0.00	0	0	0	0

Table 4.A1.5. **Environmental assessment of historical policy reforms: Hilly region** (*cont.*)

Variable	Region	Unit	RP 93-98 Actual	RP 99-03 Actual	RP 04-07 Actual	RP 08-12 Actual	RP 93-98 No-reform	RP 99-03 No-reform	RP 04-07 No-reform	RP 08-12 No-reform	RP 93-98 Change	RP 99-03 Change	RP 04-07 Change	RP 08-12 Change	RP 93-98 % change	RP 99-03 % change	RP 04-07 % change	RP 08-12 % change
P manure (total)	chh	1 000 tons P	7.64	7.27	7.47	7.54	7.76	7.47	7.13	7.27	-0.11	-0.20	0.33	0.26	-1	-3	5	4
P inputs	chh	1 000 tons P	9.37	8.56	8.61	8.63	9.65	9.25	8.36	8.46	-0.28	-0.70	0.25	0.17	-3	-8	3	2
P uptakes (main crops)	chh	1 000 tons P	0.66	0.52	0.50	0.46	0.86	0.97	0.60	0.58	-0.20	-0.45	-0.10	-0.12	-24	-46	-16	-21
P uptakes (other land)	chh	1 000 tons P	6.55	6.69	6.75	6.79	6.59	6.66	6.73	6.79	-0.03	0.03	0.01	0.00	0	0	0	0
P uptakes (total)	chh	1 000 tons P	7.21	7.21	7.25	7.25	7.45	7.63	7.33	7.37	-0.23	-0.42	-0.08	-0.12	-3	-6	-1	-2
Gross Phosphorous Balance	chh	1 000 tons P	2.16	1.35	1.37	1.38	2.21	1.62	1.03	1.08	-0.05	-0.28	0.34	0.29	-2	-17	33	27
Gross Phosphorous Balance per hectare	chh	Kg P/ha	7.73	4.85	4.96	5.04	7.89	5.85	3.74	3.97	-0.16	-1.00	1.22	1.07	-2	-17	33	27
CH_4 enteric fermentation (dairy)	chh	1 000 tons CH_4	27.12	25.82	25.48	26.10	27.31	29.02	24.65	25.06	-0.20	-3.20	0.83	1.04	-1	-11	3	4
CH_4 management (dairy)	chh	1 000 tons CH_4	5.15	4.91	4.95	5.12	5.19	5.52	4.79	4.91	-0.04	-0.61	0.16	0.20	-1	-11	3	4
CH_4 emissions (dairy)	chh	1 000 tons CH_4	32.26	30.73	30.43	31.22	32.50	34.54	29.44	29.97	-0.24	-3.81	0.99	1.25	-1	-11	3	4
CH_4 enteric fermentation (viande)	chh	1 000 tons CH_4	10.27	10.21	10.82	11.63	10.88	8.94	9.30	10.84	-0.60	1.27	1.51	0.79	-6	14	16	7
CH_4 manure management (beef)	chh	1 000 tons CH_4	1.27	1.24	1.36	1.50	1.35	1.08	1.17	1.40	-0.07	0.15	0.19	0.10	-5	14	16	7
CH_4 emissions (beef)	chh	1 000 tons CH_4	11.55	11.45	12.18	13.13	12.22	10.02	10.47	12.24	-0.68	1.43	1.71	0.89	-6	14	16	7
CH_4 emissions (cattle)	chh	1 000 tons CH_4	43.81	42.18	42.61	44.35	44.72	44.56	39.92	42.21	-0.91	-2.39	2.69	2.14	-2	-5	7	5
CH_4 other livestock	chh	1 000 tons CH_4	4.90	4.93	5.22	5.19	4.90	4.93	5.22	5.19	0.00	0.00	0.00	0.00	0	0	0	0
CH_4 emissions (total)	chh	1 000 tons CH_4	48.71	47.11	47.83	49.54	49.62	49.50	45.14	47.40	-0.91	-2.39	2.69	2.14	-2	-5	6	5
N_2O manure management	chh	1 000 tons N_2O	0.40	0.34	0.32	0.33	0.41	0.35	0.31	0.31	-0.01	-0.01	0.02	0.01	-2	-2	6	4
N_2O direct soil emissions	chh	1 000 tons N_2O	1.03	0.97	0.96	0.97	1.08	1.05	0.95	0.96	-0.04	-0.08	0.01	0.00	-4	-8	1	0
N_2O pasture range and paddock	chh	1 000 tons N_2O	0.15	0.23	0.25	0.26	0.16	0.24	0.24	0.25	0.00	-0.01	0.01	0.01	-2	-4	6	4
N_2O athmospheric deposition	chh	1 000 tons N_2O	0.08	0.08	0.08	0.08	0.08	0.08	0.08	0.08	0.00	0.00	0.00	0.00	0	0	0	0
N_2O leaching and run-off	chh	1 000 tons N_2O	0.43	0.40	0.40	0.41	0.45	0.44	0.39	0.40	-0.02	-0.04	0.01	0.01	-4	-8	4	2
N_2O emissions (total)	chh	1 000 tons N_2O	2.10	2.03	2.02	2.03	2.17	2.16	1.96	2.00	-0.07	-0.13	0.06	0.04	-3	-6	3	2
Global warming potential	chh	1 000 tons eqCO_2	1 673.59	1 617.30	1 630.97	1 670.84	1 714.20	1 708.97	1 556.04	1 614.86	-40.61	-91.66	74.93	55.98	-2	-5	5	3

Source: Model simulation with OECD Policy Evaluation Model.

Table 4.A1.6. **Environmental assessment of AP 14-17 and further reforms: Hilly region**

Variable	Region	Unit	2012 Actual	AP 14-17 Reform	EU market integration Reform	With compensation Reform	AP 14-17 Change	EU market integration Change	With compensation Change	AP 14-17 % change	EU market integration % change	With compensation % change
Chemical quantities (main crops)	chh	1 000 tons a.i.	0.04	0.04	0.05	0.05	0.00	-0.01	-0.01	0	24	19
N synthetic fertilizer quantities (main crops)	chh	1 000 tons N	2.32	2.33	2.87	2.76	0.00	-0.55	-0.44	0	24	19
N synthetic fertilizer quantities (other land)	chh	1 000 tons N	6.74	6.74	7.02	6.97	0.00	-0.29	-0.23	0	4	3
N synthetic fertilizer quantities (total)	chh	1 000 tons N	9.06	9.07	9.89	9.73	-0.01	-0.83	-0.67	0	9	7
N biological fixation	chh	1 000 tons N	10.14	10.14	9.61	9.74	0.00	0.53	0.40	0	-5	-4
N atmospheric deposition	chh	1 000 tons N	4.98	4.98	4.93	4.95	0.00	0.05	0.03	0	-1	-1
N manure from dairy cattle	chh	1 000 tons N	21.90	19.87	20.96	21.31	2.03	0.93	0.59	-9	-4	-3
N manure from beef cattle	chh	1 000 tons N	11.32	8.24	10.64	11.15	3.08	0.68	0.17	-27	-6	-1
N manure from other livestock	chh	1 000 tons N	9.04	9.04	9.04	9.04	0.00	0.00	0.00	0	0	0
N manure (total)	chh	1 000 tons N	42.26	37.15	40.65	41.50	5.11	1.61	0.76	-12	-4	-2
N inputs	chh	1 000 tons N	66.44	61.34	65.08	65.92	5.10	1.36	0.51	-8	-2	-1
N uptakes (main crops)	chh	1 000 tons N	2.24	2.25	2.76	2.66	-0.01	-0.53	-0.42	0	24	19
N uptakes (other land)	chh	1 000 tons N	39.57	39.56	38.32	38.66	0.00	1.25	0.91	0	-3	-2
N uptakes (total)	chh	1 000 tons N	41.80	41.81	41.08	41.32	-0.01	0.72	0.49	0	-2	-1
Gross Nitrogen Balance	chh	1 000 tons N	24.64	19.53	23.99	24.61	5.11	0.64	0.03	-21	-3	0
Gross Nitrogen Balance per hectare	chh	Kg N/ha	90.47	71.71	88.10	90.37	18.76	2.37	0.10	-21	-3	0
P synthetic fertilizer quantities (main crops)	chh	1 000 tons P	0.39	0.39	0.48	0.46	0.00	-0.09	-0.07	0	24	19
P synthetic fertilizer quantities (other land)	chh	1 000 tons P	0.62	0.62	0.78	0.74	0.00	-0.16	-0.13	0	26	20
P synthetic fertilizer quantities (total)	chh	1 000 tons P	1.01	1.01	1.26	1.20	0.00	-0.25	-0.20	0	25	20
P atmospheric deposition	chh	1 000 tons P	0.07	0.07	0.07	0.07	0.00	0.00	0.00	0	-1	-1
P manure from dairy cattle	chh	1 000 tons P	3.23	2.92	3.09	3.14	0.31	0.13	0.09	-10	-4	-3
P manure from beef cattle	chh	1 000 tons P	1.66	1.20	1.56	1.63	0.45	0.10	0.02	-27	-6	-1
P manure from other livestock	chh	1 000 tons P	2.60	2.60	2.60	2.60	0.00	0.00	0.00	0	0	0

Table 4.A1.6. **Environmental assessment of AP 14-17 and further reforms: Hilly region** *(cont.)*

Variable	Region	Unit	2012 Actual	AP 14-17 Reform	EU market integration Reform	With compensation Reform	AP 14-17 Change	EU market integration Change	With compensation Change	AP 14-17 % change	EU market integration % change	With compensation % change
P manure (total)	chh	1 000 tons P	7.48	6.72	7.25	7.37	0.76	0.23	0.11	-10	-3	-1
P inputs	chh	1 000 tons P	8.56	7.80	8.58	8.65	0.76	-0.02	-0.09	-9	0	1
P uptakes (main crops)	chh	1 000 tons P	0.44	0.44	0.55	0.53	0.00	-0.11	-0.08	0	24	19
P uptakes (other land)	chh	1 000 tons P	6.78	6.78	6.58	6.63	0.00	0.21	0.15	0	-3	-2
P uptakes (total)	chh	1 000 tons P	7.23	7.23	7.13	7.16	0.00	0.10	0.07	0	-1	-1
Gross Phosphorous Balance	chh	1 000 tons P	1.33	0.57	1.45	1.49	0.76	-0.12	-0.15	-57	9	12
Gross Phosphorous Balance per hectare	chh	Kg P/ha	4.90	2.11	5.33	5.47	2.79	-0.43	-0.57	-57	9	12
CH_4 enteric fermentation (dairy)	chh	1 000 tons CH_4	26.09	23.98	24.90	25.35	2.11	1.19	0.75	-8	-5	-3
CH_4 management (dairy)	chh	1 000 tons CH_4	5.12	4.70	4.88	4.97	0.41	0.23	0.15	-8	-5	-3
CH_4 emissions (dairy)	chh	1 000 tons CH_4	31.21	28.68	29.78	30.32	2.53	1.43	0.89	-8	-5	-3
CH_4 enteric fermentation (viande)	chh	1 000 tons CH_4	11.53	8.39	10.84	11.36	3.14	0.69	0.17	-27	-6	-1
CH_4 manure management (beef)	chh	1 000 tons CH_4	1.50	1.09	1.41	1.47	0.41	0.09	0.02	-27	-6	-1
CH_4 emissions (beef)	chh	1 000 tons CH_4	13.03	9.48	12.25	12.84	3.55	0.78	0.19	-27	-6	-1
CH_4 emissions (cattle)	chh	1 000 tons CH_4	44.24	38.16	42.03	43.16	6.08	2.21	1.08	-14	-5	-2
CH_4 other livestock	chh	1 000 tons CH_4	5.16	5.16	5.16	5.16	0.00	0.00	0.00	0	0	0
CH_4 emissions (total)	chh	1 000 tons CH_4	49.40	43.32	47.19	48.32	6.08	2.21	1.08	-12	-4	-2
N_2O manure management	chh	1 000 tons N_2O	0.32	0.29	0.31	0.32	0.04	0.01	0.00	-12	-3	-1
N_2O direct soil emissions	chh	1 000 tons N_2O	0.96	0.91	0.95	0.96	0.05	0.01	0.00	-5	-1	0
N_2O pasture range and paddock	chh	1 000 tons N_2O	0.25	0.22	0.24	0.25	0.03	0.01	0.00	-12	-4	-2
N_2O athmospheric deposition	chh	1 000 tons N_2O	0.08	0.08	0.08	0.08	0.00	0.00	0.00	0	-1	-1
N_2O leaching and run-off	chh	1 000 tons N_2O	0.40	0.36	0.40	0.40	0.04	0.01	0.00	-10	-2	0
N_2O emissions (total)	chh	1 000 tons N_2O	2.02	1.86	1.99	2.01	0.16	0.04	0.01	-8	-2	-1
Global warming potential	chh	1 000 tons eqCO_2	1 664.79	1 487.08	1 606.39	1 638.06	177.71	58.40	26.74	-11	-4	-2

Source: Model simulation with OECD Policy Evaluation Model.

Table 4.A1.7. Environmental assessment of historical policy reforms: Mountain region

Variable	Region	Unit	RP 93-98 Actual	RP 99-03 Actual	RP 04-07 Actual	RP 08-12 Actual	RP 93-98 No-reform	RP 99-03 No-reform	RP 04-07 No-reform	RP 08-12 No-reform	RP 93-98 Change	RP 99-03 Change	RP 04-07 Change	RP 08-12 Change	RP 93-98 % change	RP 99-03 % change	RP 04-07 % change	RP 08-12 % change
Chemical quantities (main crops)	chm	1 000 tons a.i.	0.01	0.00	0.00	0.00	0.01	0.01	0.00	0.01	0.00	0.00	0.00	0.00	-23	-42	6	-28
N synthetic fertilizer quantities (main crops)	chm	1 000 tons N	0.50	0.30	0.24	0.19	0.65	0.52	0.23	0.27	-0.15	-0.22	0.00	-0.08	-23	-43	2	-29
N synthetic fertilizer quantities (other land)	chm	1 000 tons N	1.05	1.30	1.30	1.32	1.02	1.31	1.31	1.32	0.03	-0.01	0.00	0.00	3	0	0	0
N synthetic fertilizer quantities (total)	chm	1 000 tons N	1.55	1.60	1.54	1.51	1.67	1.83	1.54	1.59	-0.12	-0.23	0.00	-0.08	-7	-13	0	-5
N biological fixation	chm	1 000 tons N	6.06	6.27	6.26	6.15	6.08	6.27	6.26	6.15	-0.02	0.01	0.00	0.00	0	0	0	0
N atmospheric deposition	chm	1 000 tons N	5.15	5.25	5.25	5.21	5.14	5.25	5.25	5.21	0.00	0.00	0.00	0.00	0	0	0	0
N manure from dairy cattle	chm	1 000 tons N	16.35	15.61	15.42	15.78	16.44	17.29	14.83	13.86	-0.10	-1.69	0.59	1.92	-1	-10	4	14
N manure from beef cattle	chm	1 000 tons N	9.43	9.65	10.32	10.95	9.68	8.48	9.44	9.02	-0.25	1.16	0.88	1.93	-3	14	9	21
N manure from other livestock	chm	1 000 tons N	5.13	5.85	5.74	5.59	5.13	5.85	5.74	5.59	0.00	0.00	0.00	0.00	0	0	0	0
N manure (total)	chm	1 000 tons N	30.91	31.10	31.49	32.33	31.25	31.62	30.01	28.47	-0.34	-0.52	1.48	3.85	-1	-2	5	14
N inputs	chm	1 000 tons N	43.66	44.23	44.54	45.20	44.15	44.97	43.06	41.42	-0.48	-0.75	1.48	3.78	-1	-2	3	9
N uptakes (main crops)	chm	1 000 tons N	0.30	0.22	0.18	0.15	0.40	0.39	0.19	0.22	-0.10	-0.17	-0.01	-0.06	-24	-43	-7	-29
N uptakes (other land)	chm	1 000 tons N	25.21	25.77	25.74	25.32	25.29	25.75	25.74	25.31	-0.08	0.02	0.01	0.01	0	0	0	0
N uptakes (total)	chm	1 000 tons N	25.51	25.99	25.92	25.47	25.69	26.14	25.93	25.53	-0.18	-0.15	-0.01	-0.06	-1	-1	0	0
Gross Nitrogen Balance	chm	1 000 tons N	18.15	18.24	18.62	19.72	18.46	18.83	17.13	15.89	-0.31	-0.59	1.49	3.83	-2	-3	9	24
Gross Nitrogen Balance per hectare	chm	Kg N/ha	63.55	62.53	63.80	68.14	64.61	64.55	58.70	54.90	-1.05	-2.02	5.10	13.24	-2	-3	9	24
P synthetic fertilizer quantities (main crops)	chm	1 000 tons P	0.08	0.05	0.04	0.03	0.11	0.09	0.04	0.04	-0.02	-0.04	0.00	-0.01	-23	-43	2	-29
P synthetic fertilizer quantities (other land)	chm	1 000 tons P	0.11	0.10	0.09	0.10	0.11	0.10	0.09	0.10	0.01	0.00	0.00	0.00	8	-2	-1	-1
P synthetic fertilizer quantities (total)	chm	1 000 tons P	0.20	0.15	0.13	0.13	0.21	0.19	0.13	0.14	-0.02	-0.04	0.00	-0.01	-7	-20	0	-10
P atmospheric deposition	chm	1 000 tons P	0.10	0.09	0.08	0.08	0.10	0.09	0.08	0.08	0.00	0.00	0.00	0.00	0	0	0	0
P manure from dairy cattle	chm	1 000 tons P	2.92	2.64	2.58	2.50	2.93	2.88	2.49	2.17	-0.01	-0.24	0.09	0.34	0	-8	4	16
P manure from beef cattle	chm	1 000 tons P	1.37	1.42	1.52	1.60	1.41	1.25	1.39	1.32	-0.04	0.17	0.13	0.28	-3	14	9	21
P manure from other livestock	chm	1 000 tons P	2.11	2.27	2.25	2.16	2.11	2.27	2.25	2.16	0.00	0.00	0.00	0.00	0	0	0	0

Table 4.A1.7. Environmental assessment of historical policy reforms: Mountain region (*cont.*)

Variable	Region	Unit	RP 93-98 Actual	RP 99-03 Actual	RP 04-07 Actual	RP 08-12 Actual	RP 93-98 No-reform	RP 99-03 No-reform	RP 04-07 No-reform	RP 08-12 No-reform	RP 93-98 Change	RP 99-03 Change	RP 04-07 Change	RP 08-12 Change	RP 93-98 % change	RP 99-03 % change	RP 04-07 % change	RP 08-12 % change
P manure (total)	chm	1 000 tons P	6.40	6.34	6.35	6.27	6.45	6.41	6.13	5.65	-0.05	-0.07	0.22	0.62	-1	-1	4	11
P inputs	chm	1 000 tons P	6.70	6.57	6.57	6.48	6.76	6.68	6.34	5.87	-0.07	-0.11	0.22	0.61	-1	-2	4	10
P uptakes (main crops)	chm	1 000 tons P	0.07	0.05	0.04	0.03	0.09	0.08	0.04	0.05	-0.02	-0.04	0.00	-0.01	-25	-43	-7	-30
P uptakes (other land)	chm	1 000 tons P	4.49	4.63	4.64	4.61	4.50	4.63	4.64	4.61	-0.01	0.00	0.00	0.00	0	0	0	0
P uptakes (total)	chm	1 000 tons P	4.56	4.68	4.68	4.65	4.59	4.71	4.68	4.66	-0.04	-0.03	0.00	-0.01	-1	-1	0	0
Gross Phosphorous Balance	chm	1 000 tons P	2.14	1.90	1.89	1.83	2.17	1.97	1.66	1.21	-0.03	-0.07	0.23	0.62	-1	-4	14	51
Gross Phosphorous Balance per hectare	chm	Kg P/ha	7.50	6.51	6.46	6.33	7.60	6.76	5.68	4.19	-0.10	-0.25	0.77	2.14	-1	-4	14	51
CH_4 enteric fermentation (dairy)	chm	1 000 tons CH_4	18.86	17.80	17.49	17.75	18.97	19.69	16.82	15.56	-0.11	-1.89	0.67	2.19	-1	-10	4	14
CH_4 management (dairy)	chm	1 000 tons CH_4	3.58	3.38	3.40	3.48	3.60	3.74	3.27	3.05	-0.02	-0.36	0.13	0.43	-1	-10	4	14
CH_4 emissions (dairy)	chm	1 000 tons CH_4	22.44	21.19	20.89	21.23	22.57	23.43	20.09	18.62	-0.13	-2.24	0.80	2.61	-1	-10	4	14
CH_4 enteric fermentation (viande)	chm	1 000 tons CH_4	9.67	10.07	10.57	11.19	9.92	8.85	9.66	9.22	-0.25	1.22	0.90	1.98	-2	14	9	21
CH_4 manure management (beef)	chm	1 000 tons CH_4	1.20	1.22	1.33	1.45	1.23	1.08	1.22	1.20	-0.03	0.15	0.11	0.26	-2	14	9	21
CH_4 emissions (beef)	chm	1 000 tons CH_4	10.87	11.29	11.90	12.65	11.14	9.93	10.88	10.41	-0.28	1.36	1.02	2.23	-2	14	9	21
CH_4 emissions (cattle)	chm	1 000 tons CH_4	33.30	32.48	32.79	33.88	33.71	33.36	30.97	29.03	-0.40	-0.88	1.82	4.85	-1	-3	6	17
CH_4 other livestock	chm	1 000 tons CH_4	3.93	4.05	4.01	3.90	3.93	4.05	4.01	3.90	0.00	0.00	0.00	0.00	0	0	0	0
CH_4 emissions (total)	chm	1 000 tons CH_4	37.23	36.53	36.81	37.78	37.64	37.41	34.99	32.93	-0.40	-0.88	1.82	4.85	-1	-2	5	15
N_2O manure management	chm	1 000 tons N_2O	0.32	0.30	0.28	0.28	0.32	0.30	0.27	0.25	0.00	0.00	0.01	0.03	-1	0	4	11
N_2O direct soil emissions	chm	1 000 tons N_2O	0.64	0.64	0.64	0.64	0.65	0.65	0.62	0.60	-0.01	-0.01	0.02	0.04	-1	-2	2	6
N_2O pasture range and paddock	chm	1 000 tons N_2O	0.11	0.17	0.19	0.19	0.11	0.17	0.18	0.17	0.00	0.00	0.01	0.02	-1	-2	5	14
N_2O athmospheric deposition	chm	1 000 tons N_2O	0.08	0.08	0.08	0.08	0.08	0.08	0.08	0.08	0.00	0.00	0.00	0.00	0	0	0	0
N_2O leaching and run-off	chm	1 000 tons N_2O	0.26	0.26	0.26	0.27	0.26	0.26	0.25	0.24	0.00	-0.01	0.01	0.03	-1	-2	5	13
N_2O emissions (total)	chm	1 000 tons N_2O	1.41	1.44	1.44	1.46	1.42	1.46	1.40	1.34	-0.02	-0.02	0.05	0.12	-1	-1	3	9
Global warming potential	chm	1 000 tons eqCO_2	1 218.18	1 214.35	1 220.10	1 245.81	1 231.73	1 238.64	1 167.26	1 107.41	-13.55	-24.29	52.84	138.39	-1	-2	5	12

Source: Model simulation with OECD Policy Evaluation Model.

Table 4.A1.8. Environmental assessment of AP 14-17 and further reforms: Mountain region

Variable	Region	Unit	2012 Actual	AP 14-17 Reform	EU market integration Reform	With compensation Reform	AP 14-17 Change	EU market integration Change	With compensation Change	AP 14-17 % change	EU market integration % change	With compensation % change
Chemical quantities (main crops)	chm	1 000 tons a.i.	0.00	0.00	0.00	0.00	0.00	0.00	0.00	9	28	4
N synthetic fertilizer quantities (main crops)	chm	1 000 tons N	0.18	0.20	0.23	0.19	-0.02	-0.05	-0.01	9	28	4
N synthetic fertilizer quantities (other land)	chm	1 000 tons N	1.33	1.36	1.41	1.34	-0.03	-0.08	-0.01	3	6	1
N synthetic fertilizer quantities (total)	chm	1 000 tons N	1.51	1.56	1.64	1.53	-0.05	-0.13	-0.02	3	9	1
N biological fixation	chm	1 000 tons N	6.12	6.09	6.04	6.11	0.03	0.07	0.01	-1	-1	0
N atmospheric deposition	chm	1 000 tons N	5.22	5.21	5.19	5.21	0.01	0.03	0.00	0	-1	0
N manure from dairy cattle	chm	1 000 tons N	15.82	14.88	14.58	14.47	0.93	1.23	1.35	-6	-8	-9
N manure from beef cattle	chm	1 000 tons N	10.83	9.06	10.53	11.13	1.77	0.31	-0.30	-16	-3	3
N manure from other livestock	chm	1 000 tons N	5.45	5.45	5.45	5.45	0.00	0.00	0.00	0	0	0
N manure (total)	chm	1 000 tons N	32.10	29.39	30.56	31.05	2.71	1.54	1.05	-8	-5	-3
N inputs	chm	1 000 tons N	44.94	42.24	43.43	43.90	2.70	1.51	1.04	-6	-3	-2
N uptakes (main crops)	chm	1 000 tons N	0.15	0.16	0.19	0.15	-0.01	-0.04	-0.01	9	28	4
N uptakes (other land)	chm	1 000 tons N	25.17	25.08	24.97	25.14	0.09	0.19	0.02	0	-1	0
N uptakes (total)	chm	1 000 tons N	25.31	25.24	25.16	25.30	0.08	0.15	0.02	0	-1	0
Gross Nitrogen Balance	chm	1 000 tons N	19.63	17.01	18.27	18.60	2.62	1.36	1.02	-13	-7	-5
Gross Nitrogen Balance per hectare	chm	Kg N/ha	67.74	58.69	63.05	64.21	9.04	4.68	3.52	-13	-7	-5
P synthetic fertilizer quantities (main crops)	chm	1 000 tons P	0.03	0.03	0.04	0.03	0.00	-0.01	0.00	9	28	4
P synthetic fertilizer quantities (other land)	chm	1 000 tons P	0.10	0.11	0.12	0.10	-0.01	-0.02	0.00	9	21	3
P synthetic fertilizer quantities (total)	chm	1 000 tons P	0.13	0.14	0.16	0.13	-0.01	-0.03	0.00	9	23	3
P atmospheric deposition	chm	1 000 tons P	0.08	0.08	0.08	0.08	0.00	0.00	0.00	0	-1	0
P manure from dairy cattle	chm	1 000 tons P	2.49	2.26	2.33	2.32	0.23	0.16	0.17	-9	-6	-7
P manure from beef cattle	chm	1 000 tons P	1.58	1.33	1.54	1.63	0.26	0.04	-0.04	-16	-3	3
P manure from other livestock	chm	1 000 tons P	2.08	2.08	2.08	2.08	0.00	0.00	0.00	0	0	0

Table 4.A1.8. **Environmental assessment of AP 14-17 and further reforms: Mountain region** *(cont.)*

Variable	Region	Unit	2012 Actual	AP 14-17 Reform	EU market integration Reform	With compensation Reform	AP 14-17 Change	EU market integration Change	With compensation Change	AP 14-17 % change	EU market integration % change	With compensation % change
P manure (total)	chm	1 000 tons P	6.16	5.67	5.95	6.03	0.49	0.20	0.13	-8	-3	-2
P inputs	chm	1 000 tons P	6.37	5.89	6.19	6.24	0.48	0.17	0.12	-7	-3	-2
P uptakes (main crops)	chm	1 000 tons P	0.03	0.03	0.04	0.03	0.00	-0.01	0.00	9	28	4
P uptakes (other land)	chm	1 000 tons P	4.62	4.60	4.58	4.61	0.02	0.04	0.00	0	-1	0
P uptakes (total)	chm	1 000 tons P	4.65	4.64	4.62	4.65	0.01	0.03	0.00	0	-1	0
Gross Phosphorous Balance	chm	1 000 tons P	1.72	1.25	1.57	1.59	0.46	0.14	0.12	-27	-8	-7
Gross Phosphorous Balance per hectare	chm	Kg P/ha	5.92	4.33	5.42	5.50	1.59	0.50	0.42	-27	-8	-7
CH_4 enteric fermentation (dairy)	chm	1 000 tons CH_4	17.75	16.64	16.40	16.27	1.11	1.36	1.48	-6	-8	-8
CH_4 management (dairy)	chm	1 000 tons CH_4	3.48	3.26	3.21	3.19	0.22	0.27	0.29	-6	-8	-8
CH_4 emissions (dairy)	chm	1 000 tons CH_4	21.23	19.91	19.61	19.46	1.33	1.63	1.77	-6	-8	-8
CH_4 enteric fermentation (viande)	chm	1 000 tons CH_4	11.15	9.32	10.83	11.46	1.82	0.32	-0.31	-16	-3	3
CH4 manure management (beef)	chm	1 000 tons CH_4	1.45	1.22	1.41	1.49	0.24	0.04	-0.04	-16	-3	3
CH_4 emissions (beef)	chm	1 000 tons CH_4	12.60	10.54	12.24	12.95	2.06	0.36	-0.35	-16	-3	3
CH_4 emissions (cattle)	chm	1 000 tons CH_4	33.83	30.44	31.85	32.41	3.39	1.98	1.42	-10	-6	-4
CH_4 other livestock	chm	1 000 tons CH_4	3.83	3.83	3.83	3.83	0.00	0.00	0.00	0	0	0
CH_4 emissions (total)	chm	1 000 tons CH_4	37.66	34.27	35.68	36.24	3.39	1.98	1.42	-9	-5	-4
N_2O manure management	chm	1 000 tons N_2O	0.27	0.25	0.26	0.27	0.02	0.01	0.00	-8	-3	-2
N_2O direct soil emissions	chm	1 000 tons N_2O	0.64	0.18	0.18	0.19	0.46	0.45	0.45	-72	-71	-71
N_2O pasture range and paddock	chm	1 000 tons N_2O	0.19	0.18	0.18	0.19	0.02	0.01	0.01	-8	-5	-3
N_2O atmospheric deposition	chm	1 000 tons N_2O	0.08	0.08	0.08	0.08	0.00	0.00	0.00	0	-1	0
N_2O leaching and run-off	chm	1 000 tons N_2O	0.26	0.24	0.25	0.26	0.02	0.01	0.01	-8	-4	-3
N_2O emissions (total)	chm	1 000 tons N_2O	1.45	0.93	0.97	0.98	0.52	0.48	0.47	-36	-33	-32
Global warming potential	chm	1 000 tons eqCO_2	1 240.71	1 142.72	1 184.64	1 201.64	98.00	56.07	39.07	-8	-5	-3

Source: Model simulation with OECD Policy Evaluation Model.

Chapter 5

Competitiveness of Swiss food industries*

This chapter evaluates strength and weaknesses of the Swiss food industries and their competitiveness on domestic and EU markets. The competitiveness of the industry is assessed by several industry indicators such as global turnover, labour productivity and international trade indicators. Selected EU countries are used as a benchmark to assess the competitiveness of the Swiss agro-food sector as a whole and for selected agro-food branches.

* This chapter presents some of the material provided in a consultant study prepared for this report by WUR-LEI.

Objective and methodology

This chapter presents an ex-post assessment of the competiveness performance of Swiss food industries. It benchmarks the industry against is main competitors in the EU and provides information on the structure of the Swiss food industry and its raw materials base.[1]

Although competitiveness does not have a precise definition in economic theory it can be understood as the ability to successfully face competition. In this sense, competitiveness is the ability to sell products that meet demand requirements (price, quality, quantity) and, at the same time, ensure profits over time that enable the firm to thrive. Competitiveness is a relative concept and should be measured with respect to a benchmark. Competition may be within domestic markets (in which case firms within one sector, or entire sectors in the same country are compared with each other) or international (in this case, comparisons are made between countries).

The approach taken here is to measure revealed performance, relying on indicators such as market performance, trade success and revealed comparative advantage indicators. (Wijnands et al., 2007; Latruffe, 2010). The analysis of competitiveness examines the ex-post performance of an industry in Switzerland compared to the same industry in benchmark countries. The selected indicators to quantify competitiveness are:[2]

Trade related indicators:

- Growth (as difference between 2 periods) of the export share on the world market of a specific subsector of the food industry or of the food industry as whole. The market share of one country is compared with the total world export of that (sub-) industry. This performance indicator reflects the outcome of the competitive process on international markets.
- The difference of the Relative Trade Advantage (RTA) index between 2 periods. The RTA is defined by Scott and Vollrath (1992) as the difference between the Relative Export Advantage index (RXA) and the Relative Import Advantage index (RMA). A positive RTA indicates a competitive advantage: exports exceed imports. Negative values signify competitive disadvantages.[3]

Economic performance indicators:

- Annual growth of real turnover of a specific industry relative to the total food industry. This indicator reflects the competition for factors of production between different industries within a country. Ideally, growth of industry value added relative to that of the total food industry would have been used to construct this indicator, but this information was not available;
- Annual growth of the real turnover per employee as an indicator for labour productivity. This affects the unit labour costs and in this way relative prices.
- Annual growth of turnover reflects the performance of that specific (sub-) industry.

As competitiveness is a relative concept, the performance of the Swiss food industries is compared with the food industries in benchmark countries. Because the EU is by far its main market the Swiss food industry is compared with selected EU countries. EU-countries that have at least a share of 5% in Swiss exports or imports of agro-food products are considered *benchmark countries*: Austria, France, Germany, Italy, the Netherlands, Spain and the United Kingdom. These countries represent 86% of the Swiss export value to the EU and 89% of Swiss import value from the EU. All data for the competitiveness assessment are taken from publicly available data sources.[4, 5]

Competitiveness of the food and beverages industry

Overall findings

Benchmarking of the Swiss food and beverage industry against its main competitors in the EU, shows that competitiveness of the Swiss food and beverage industry is almost entirely driven by sub-branches that source most of their raw material inputs abroad or where inputs are non-agricultural (mineral water). In particular, the strong sub-branch "other food" manufacturing – dominated by cocoa and chocolate manufacturing is performing very well. The turnover in this sub-branch grew annually by 10%, almost twice as fast as the overall food and beverages industry (5.8%) in the period 2001-11. The two strongest sub-sectors "other food" and beverages industry count for 72% of the exports of the Swiss agro-food industry.

On the other hand, the weakest sectors are meat and dairy, which mainly rely on domestic primary agriculture for their inputs, although some dairy producers successfully serve high value niche markets. These industries as well as the weak animal feed sector, have to pay relatively high prices for their material inputs, well above the EU price levels. Additionally, these less competitive sectors have relatively low labour productivity and are relatively labour intensive.

Given the particular importance of the chocolate industry (in the "other food" sub-sector) and to a lesser extent of soft drinks and bottled mineral waters, one can state the Swiss food competitiveness is *"cocoa and water"* based, i.e. it is not based on domestically produced agricultural raw materials.

Overall competitiveness of food and beverages

Taking together the entire food and beverages sector and aggregating all indicators of performance appears to indicate that the overall competitiveness (O) of the Swiss food industry exceeds all selected countries (Figure 5.1). This is almost entirely driven by the performance of "other food" manufacturing, which accounts for two-thirds of the turnover and half of the exports. It is thus the most important food-manufacturing sub-sector (Table 5.1). However, this sector is potentially "footloose" and only weakly linked to the domestic primary agriculture sector. The sub-sector chocolate manufacturing grew rapidly and had a share of round 50% in the turnover in 2011 in the "other food sector".

The dairy industry is the sub-sector which accounts for the most enterprises and is second in turnover, after "other food". Beverages manufacturing takes a quarter of both exports and imports of the sector. Manufacturing of other food, dairy, meat and beverages are the largest in terms of turnover and trade.

The performance of the food sector is weakly linked to primary agriculture. Most prices of primary agricultural products are above the level of the selected countries in the EU, although for several products these prices are converging to the EU level (Table 5.2).

Figure 5.1. **Competitiveness of the food & beverages industry**

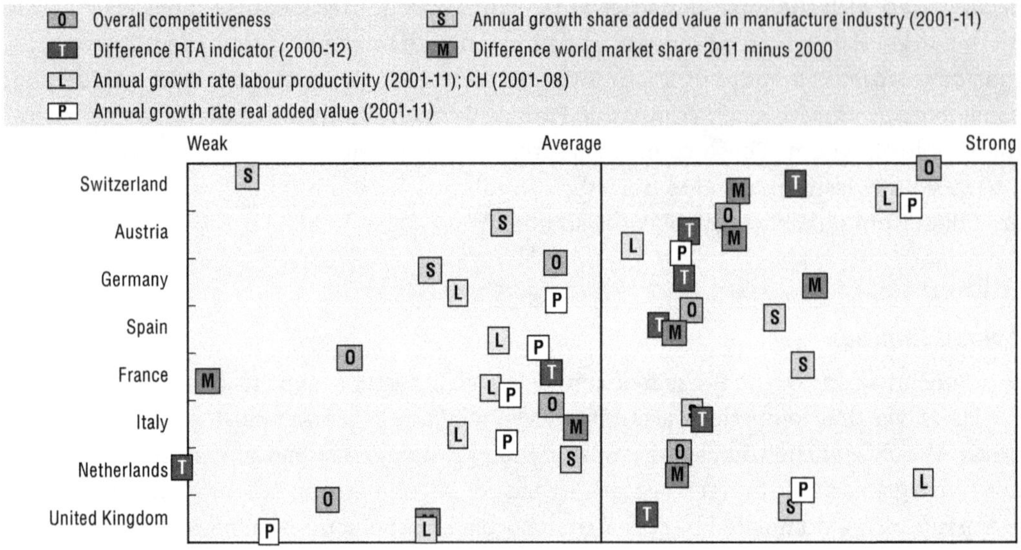

Source: LEI calculations based on Eurostat and FBS.

Table 5.1. **Key-figures of the Swiss food industry**[1]

NACE	Manufacturing industry	Entreprises 2011		Turnover 2011		Employees 2008		Export 2012		Import 2012	
		Number	%	Billion €	%	1 000	%	Million USD		Million USD	
C10&11	Food total	2 410	100	41.6	100	73.1	100	7 889	100	8 237	100
C101	Meat	257	10.7	4.1	9.9	11.9	16.3	113	1	943	11
C103	Fruit and vegetable	67	2.8	0.7	1.7	2.2	3.0	224	3	714	9
C104	Oils and fats	21	0.9	0.4	1.0	0.4	0.5	83	1	385	5
C105	Dairy	768	31.9	4.9	11.8	8.2	11.2	761	10	516	6
C106	Grain and starches	89	3.7	0.6	1.4	1.3	1.8	11	0	93	1
C107	Bakery	219	9.1	2.1	5.0	14.6	20.0	728	9	743	9
C108	Other food	471	19.5	24.4	58.7	15.1	20.7	3 896	49	1 760	21
C109	Animal feeds	142	5.9	1.3	3.1	1.9	2.6	214	3	575	7
C110	Beverages	367	15.2	3.2	7.7	7.1	9.7	1 851	23	1 901	23

1. Sum of subsector differs from Food total, because C102 "Processing and preserving of fish, crustaceans and molluscs" is not included. This industry has a substantial import: 7% of total food industry imports.

Source: LEI calculation based on FBS and UNComtrade.

Prices of pig and cattle meat are however diverging from the EU level. Self-sufficiency for most processed products is below 100% and Switzerland is a net-importer. It is a strong exporter only for "other food" products and a small exporter for dairy products.

The Swiss food industry is one of the smallest of all selected countries in terms of turnover and number of enterprises. However, in turnover it is more than two times bigger than the Austrian food industry, while both countries have a comparable population size. Swiss agro-food enterprises out-perform their EU competitors in terms of growth of turnover and turnover per enterprise (Table 5.3).

The distribution of firm size is skewed in all countries, and Switzerland is no exception: the largest 3% account for 60% of the turnover and the largest 13% account for more than 80%. Market concentration, as measured by the skewedness of the firm size distribution is observed in all Swiss food subsectors.

Table 5.2. **Switzerland: Overview of the raw material supply and self-sufficiency**

Product	Production raw materials	Price raw materials	Self-sufficiency
Meat	Cattle and Pig meat declining Poultry strongly increasing	All above benchmark countries Pig meat prices converging Cattle and poultry meat diverging	Not self-sufficient in meat Net-importer of meat products
Fruit and Vegetables	Little information available, relatively small production Apples and Potatoes declining	Apples above EU, below the UK level. Diverging Potatoes above EU level, converging	Net-importer of processed products Potatoes net-importing Tomatoes 20% self- sufficiency
Oils and fats	Few oilseeds produced Rapeseed slightly increasing	Above EU countries, converging	Net-importer of oils and fats
Dairy	Milk production slightly increasing	Above EU countries, converging	Net-exporter of dairy products
Grain mill and starche	Wheat is largest product Wheat production declined slightly	Above EU countries, converging Price peak in 2008 mitigated in Switzerland	Net-importer of grain mill and starches products
Bakery	See grain mill		Very small net-importer
Other food	Mainly footloose based on imported products such as cocoa		Large net-exporter Net-importer of sugar
Animal feed	No information		Net-importer, especially oil cakes
Beverages	Otherwise no specific information	Prices of grapes are far above the EU level	Small net-importer of beverages Grapes for wine self sufficient

Source: LEI assessment based on Eurostat for EU countries and BFS for Switzerland and UNComtrade data.

Table 5.3. **Structure of the food & beverages industry in 2011**

	Turnover		Enterprises		Average turnover per enterprise		Employees	
	Billion (€)	Growth[1] (%)	Number	Growth[1] (%)	Million (€)	Growth[1] (%)	1 000	Growth[1] (%)
Switzerland[2]	41.6	5.8	2 410	-1.0	17.3	6.8	73.1	0.0
Austria	19.3	4.6	3 837	-1.0	5	5.7	77.5	-0.1
Germany	180.4	2.4	32 204	-1.0	5.6	3.4	887.5	0.8
Spain	101.5	3.6	27 722	-1.3	3.7	5.0	365.9	-0.1
France	168.9	1.9	59 405	-1.2	2.8	3.1	604.4	-0.4
Italy	124.3	2.5	58 074	-1.6	2.1	4.2	433.5	0.0
Netherlands	62.9	3.7	4 477	-1.2	14.1	5.0	125.3	-2.5
United Kingdom	105.8	0.0	7 492	-0.3	14.1	0.2	376.3	-3.0

1. Annual growth rate from 2001 to 2011.
2. Swiss labour data are for 2008 and growth rate 2001-08.

Source: LEI calculation based on Eurostat for EU countries and BFS for Switzerland.

While the Swiss food industry showed strong growth of turnover, other manufacturing grew at an even higher pace: consequently the share of the food industry in manufacturing industry declined in the period 2001-11. In all other selected countries (except Germany), the share remained at the same level or increased. Secondly, labour productivity in Swiss food industry increased at an impressive rate of 6% per year. Only the Netherlands showed a slightly higher growth rate (Table 5.4).

From 2001 to 2011, the export of processed food and beverage products of Switzerland grew fast (15%) and above the world average (13%), such that the share on the world market increased. On the other hand, Swiss imports grew by less than world average resulting in a smaller share of food imports from the world market. Switzerland is a small trader, just as Austria with shares of round 1% on both the export and import world market. Germany as the largest exporter and importer has shares of 7%. France lost position: the export share decreased from 8.5% in 2000 to 6.0% in 2012. The United Kingdom is the largest net-importer and the Netherlands the largest net-exporter of the selected countries. The

Table 5.4. **Share of food industry in manufacturing and labour productivity (based on turnover)**

	Share in manufacturing turnover			Labour productivity (€ 1 000 turnover per employee)		
	2001 (%)	2011 (%)	Growth[1] (%)	2001	2011	Growth[1] (%)
Switzerland [2]	11.8	9.6	-2.0	319	482	6.1
Austria	11.0	11.1	0.2	153	191	2.2
Germany	9.6	9.2	-0.4	165	167	0.1
Spain	17.5	21.6	2.1	193	208	0.8
France	14.8	18.8	2.4	212	226	0.6
Italy	11.7	13.5	1.4	216	219	0.1
Netherlands	18.9	20.3	0.7	255	483	6.6
United Kingdom	14.3	17.9	2.3	202	195	-0.3

1. Annual growth rate from 2001 to 2011.
2. Swiss labour data are for 2008 and growth rate 2001-08.
Source: LEI calculation based on Eurostat for EU countries and BFS for Switzerland.

export-import balance is slightly negative for Switzerland and over the long term it is a net-importer. The Relative Trade Advantage for Switzerland is just below -0.2, indicating a non-specialized importer (Table 5.5 and Figure 5.2).

Table 5.5. **Trade and market shares in food and beverages products**

	Export				Import			
	Export 2012 (Million USD)	Growth 2000-11 (%)	Market share 2000 (%)	Market share 2011 (%)	Import 2012 (Million USD)	Growth 2000-11 (%)	Market share 2000 (%)	Market share 2011 (%)
Switzerland	7 889	15.2	0.6	0.9	8 237	9.2	1.1	1.0
Austria	10 069	13.9	0.9	1.1	9 248	12.1	0.9	1.1
Germany	64 446	12.5	6.4	7.1	59 075	10.2	7.4	7.1
Spain	29 024	11.1	3.1	3.0	23 520	9.7	3.2	2.9
France	53 620	7.8	8.5	6.0	43 276	9.1	5.9	5.0
Italy	32 574	10.2	4.0	3.6	32 122	9	4.7	3.9
Netherlands	59 633	11.3	6.8	6.7	41 461	12.9	3.7	4.6
United Kingdom	26 212	7.7	4.2	2.9	47 147	8.2	6.9	5.3

Source: LEI calculation based on UNComtrade.

Key findings for selected food sectors

Developments in the different sub-sectors of the Swiss food and beverages industry do not all follow the same path. This section provides more details on the main Swiss food sector branches: i) meat processing industry; ii) dairy products; iii) beverages; and iv) other food products (half of other food is cocoa and chocolate manufacturing). These four sub-sectors represent almost 90% of the turnover of the total Swiss agro-food sector.

Meat processing industry

The overall competitiveness (O) of the Swiss meat processing industry (C101) is the weakest of all selected countries (Figure 5.3). The main developments indicate that:

- The Swiss share of turnover of the meat industry in total manufacture (S) declined considerably and Switzerland's performance is the weakest observed.
- The growth of real turnover (P) of the meat industry is above average and on the same level as Austria, Spain and the Netherlands.

Figure 5.2. **Processed food and beverage trade indicators**

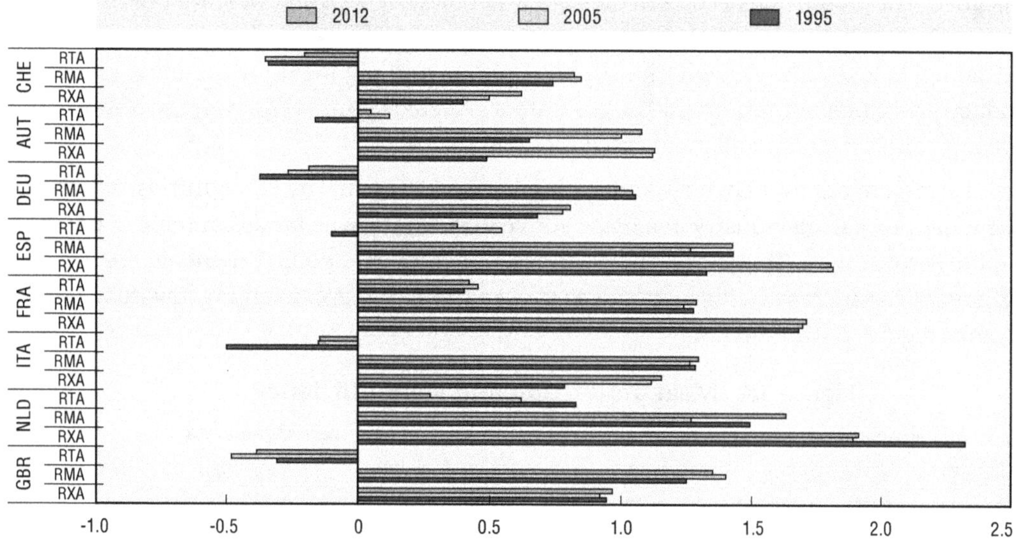

Note: RTA, RMA, RXA denote relative trade/import/export advantage indexes and are defined in Annex 5.1.
Source: LEI calculations based on UNComtrade.

- The growth of labour productivity (real turnover per employee (L) is also the weakest compared to all benchmark countries.
- The Swiss Relative Trade Advantage (T) index was stable and is above average. Switzerland remained a net-importer of meat products.
- The performance of the export share on the world market (M) of Switzerland is just below average: Switzerland gained an insignificant small market share (0.1% points). Germany gained a significant share from 5 to 9% (4% points).

Figure 5.3. **Competitiveness of the meat industry**

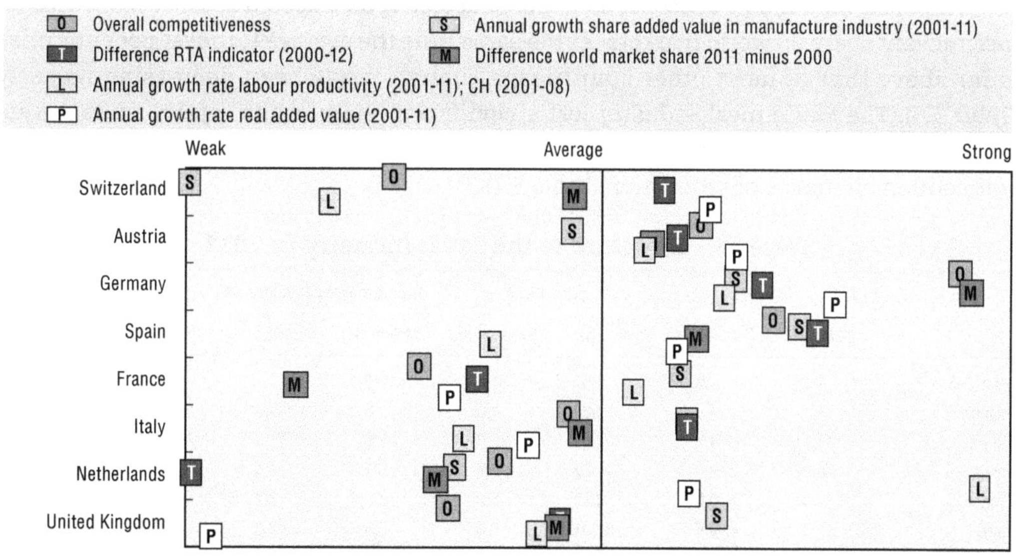

Source: LEI calculations based on Eurostat and FBS.

The supply of domestically produced cattle and pig meat to the Swiss meat industry has declined and supply of poultry meat has strongly increased in the period 1991-2011. In the period 1991-2012, the Swiss cattle (-0.9% annually) and pig meat production (-0.4%) declined. The production of chicken meat grew annually by 4.2% and was, after Germany (5%), the highest growth level. Nevertheless, Swiss self-sufficiency for all meat products remained below 100%. Overall, the supply of cattle and pig meat for the meat processing industry declined in the period 1991-2011. Most selected countries, except the Netherlands, are not self-sufficient in all meat categories (Figure 5.4).

Farm-gate prices in Switzerland are higher than those in the EU countries, which puts the meat processing industry at a competitive disadvantage as far as sourcing of domestic inputs is concerned. This is especially the case for cattle and poultry meat, where the price difference has increased. Pig meat prices show a trend that is converging towards EU levels (Annex Figures 5.A1.1-5.A1.3).

Figure 5.4. **Meat production and self-sufficiency**

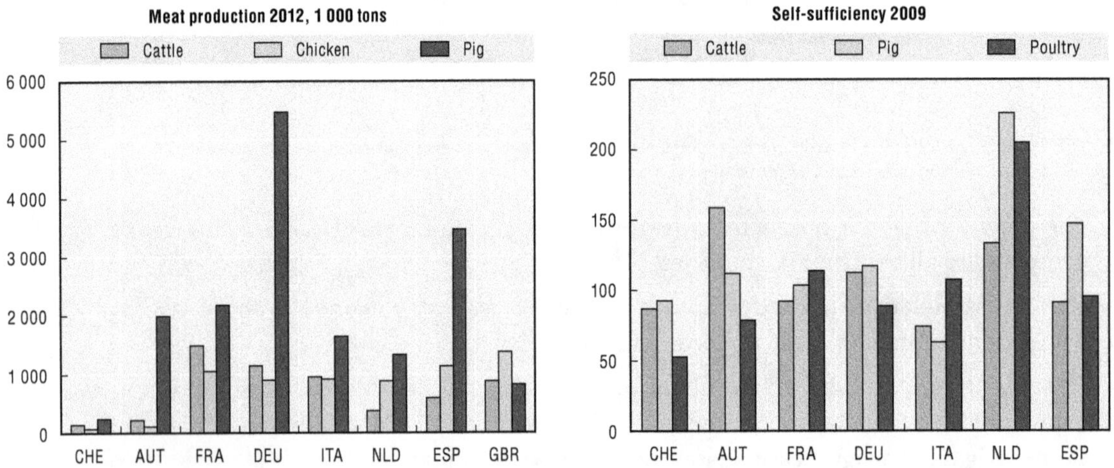

Source: LEI calculation based on FAOstat.

The meat industry in Switzerland is the smallest of all selected countries and has an insignificant share on world markets. At the same time the average turnover per enterprise is far above that of most other countries, although it has been decreasing recently (Table 5.6). The Swiss meat industry lost a significant share in the total manufacturing during 2001-11 and had decreasing labour productivity, indicating a weak position in the competition on means of production (Table 5.7).

Table 5.6. **Structure of the meat industry in 2011**

	Turnover		Enterprises		Aver. turnover per enterprise		Employees	
	Billion (€)	Growth[1] (%)	Number	Growth[1] (%)	Million (€)	Growth[1] (%)	1 000	Growth[1] (%)
Switzerland[2]	4.1	1.3	257	1.5	16.0	-0.1	11.9	-0.1
Austria	3.7	3.9	986	-1.6	3.8	5.6	16.9	-0.2
Germany	44.1	4.2	11 295	-2.6	3.9	7.0	202.1	-0.4
Spain	20.9	3.7	4 062	-0.7	5.1	4.4	83.3	1.4
France	34.9	0.0	6 540	-5.9	5.3	6.2	127.9	-3.0
Italy	19.8	1.6	3 601	-0.3	5.5	2.0	59.3	0.5
Netherlands	9.2	0.6	519	-4.5	17.8	5.4	13.9	-6.6
United Kingdom	16.9	-1.3	1 024	-1.2	16.5	0.0	74.5	-4.7

1. Annual growth rate from 2001 to 2011.
2. Swiss labour data are for 2008 and growth rate 2001-08.

Source: LEI calculation based on Eurostat for EU countries and BFS for Switzerland.

Table 5.7. **Share of meat in manufacturing and labour productivity**

Based on real turnover

	Share in manufacturing turnover			Labour productivity (€ 1 000 real turnover per employee)		
	2001 (%)	2011 (%)	Growth[1] (%)	2001	2011	Growth[1] (%)
Switzerland[2]	1.8	1.0	-6.1	294	227	-3.6
Austria	2.3	2.1	-0.6	144	169	1.6
Germany	2.0	2.3	1.3	132	179	3.1
Spain	3.6	4.4	2.2	200	188	-0.7
France	3.7	3.9	0.5	192	221	1.4
Italy	2.0	2.2	0.6	286	255	-1.1
Netherlands	3.8	3.0	-2.3	297	637	7.9
United Kingdom	2.6	2.9	1.0	154	157	0.2

1. Annual growth rate from 2001 to 2011.
2. Swiss labour data are for 2008 and growth rate 2001-08.
Source: LEI calculation based on Eurostat for EU countries and BFS for Switzerland.

Manufacture of dairy products

The overall competitiveness (O) of the Swiss dairy processing industry (C105) is weak compared to the benchmark countries (Figure 5.5). The main developments indicate that:

- The share of the turnover of the dairy industry in total manufacture (S) declined strongly and Switzerland's performance is the weakest observed.
- The growth of real turnover (P) of the dairy industry is above average.
- The growth of labour productivity (real turnover per employee (L)) is weak, especially compared to the Netherlands, but slightly better than in Austria.
- The Relative Trade Advantage (T) index of Switzerland declined; Switzerland is below average and relatively weak. Nevertheless, Switzerland remained a net-exporter of dairy products.
- The growth of the export share on the world market (M) of Switzerland is above average: the decline was less than the leading EU exporters France, Germany and the Netherlands.

Figure 5.5. **Competitiveness of the dairy industry**

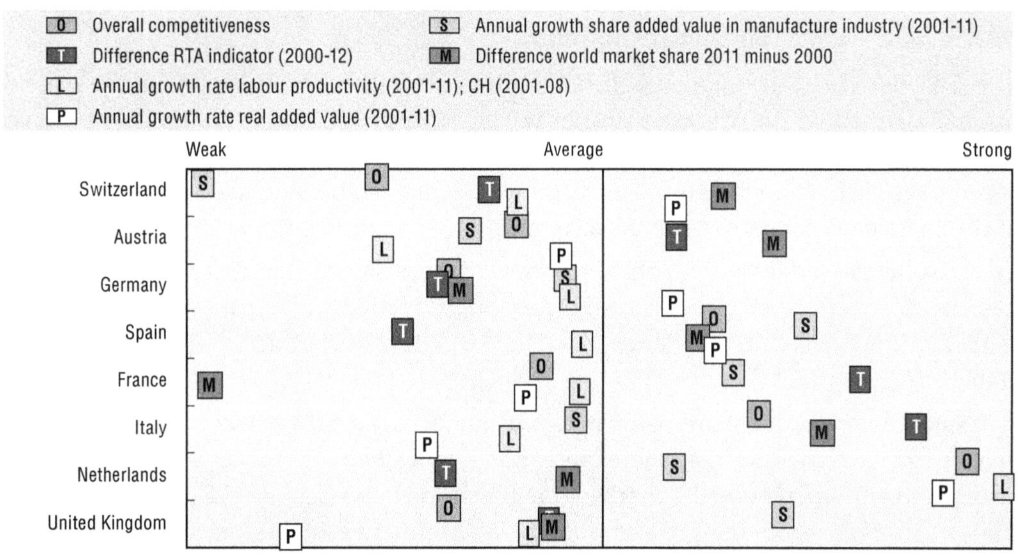

Source: LEI calculations based on Eurostat and FBS.

Milk production in Switzerland and in benchmark countries was rather stable during the period 1991 to 2009. Nevertheless a small growth for some countries including Switzerland can be observed and a decline in the milk production for some others. Self-sufficiency in Switzerland was increasing, but slowly. Countries like Austria and the Netherlands showed a stronger growth, whereas Spain and the United Kingdom showed a decline. Based on these developments we conclude that the domestic raw material base for the Swiss dairy industry did not change significantly (Table 5.8).

Table 5.8. **Production and self-sufficiency of milk**

	Production (million ton)				Self-sufficiency[1] (%)			
	2009	1991-2009			2009	1991-2009		
		Mean	Stdev	Growth		Mean	Stdev	Growth
Switzerland	4.1	3.9	0.1	0.2	117.3	113.3	2.8	0.2
Austria	3.3	3.2	0.1	-0.1	136.5	118.9	10.1	1.2
France	24.2	25.7	0.6	-0.6	127.6	124	3.8	0.4
Germany	29.2	28.4	0.4	0.0	121.3	122	5.5	-0.1
Italy	11.4	12.2	0.6	-0.3	68.8	69.4	1.9	0.1
Netherlands	11.5	11.1	0.3	0.2	163.3	135.1	12.7	0.9
Spain	7.4	7.1	0.3	0.1	70.2	80.3	6.0	-1.2
United Kingdom	13.2	14.6	0.5	-0.6	77.8	91.1	5.9	-1.1

1. Self-sufficiency is the domestic supply (= supply for domestic utilization (FAO: *http://faostat.fao.org/site/379/DesktopDefault.aspx?PageID=379*) as percentage of the production.

Source: LEI calculation based on FAOStat commodity balances: FAO item code 2848 "Milk – Excluding Butter".

Milk prices in Switzerland are still above those of selected EU countries. However, the prices are gradually converging; in the early nineties, the Swiss producer price was round 1.9 times the German price and in 2011 it was around 1.5 times. Within the EU considerable differences in prices of milk can be observed, with a 40% difference between the highest (Italy) and lowest (United Kingdom) price in 2009-11 (Annex Figure 5.A1.4).

The economic performance of the Swiss dairy industry is weak compared to the benchmark countries. The growth rate in turnover is among the lowest of the selected countries. The number of dairy processing firms declined strongly in Switzerland, stronger than in Italy, Spain or the United Kingdom.

This decline in the number of firms resulted in a rather strong growth of the scale of firms: 5% growth of the average turnover per enterprise. Still, the Swiss average turnover per enterprise is one of the lowest, together with Italy and Spain. The Dutch and German enterprises have an average turnover that is 5 to 9 times higher. In addition, employment in the Swiss dairy industry declined, as in most other countries (Table 5.9).

Despite the moderate growth of turnover, the Swiss dairy industry shows a rapidly decreasing share in total manufacturing. Labour productivity (real turnover per employee) remained unchanged in Switzerland, while the Netherlands had a very strong growth (Table 5.10).

Swiss dairy products are performing weakly on the world market compared to the benchmark countries. Market shares of exports as well as of imports are decreasing, in spite of growth in some niche markets for specialised cheeses. World trade (export and import) in dairy products grew annually by around 12% in the period 2000 to 2011. Switzerland and all selected European countries show growth rates in trade in dairy

Table 5.9. **Structure of the dairy industry in 2011**

	Turnover		Enterprises		Aver. turnover per enterprise		Employees	
	Billion (€)	Growth[1] (%)	Number	Growth[1] (%)	Million (€)	Growth[1] (%)	1 000	Growth[1] (%)
Switzerland[2]	4.9	0.8	768	-4.0	6.4	5.0	8.2	-2.2
Austria	2.4	2.1	157	3.3	15.5	-1.2	4.9	0.8
Germany	27.5	2.1	472	3.9	58.2	-1.7	40.1	0.1
Spain	10.6	4	1 445	-0.3	7.3	4.3	26.8	0.5
France	27.2	0.9	1 958	2.7	13.9	-1.7	57.2	-1.2
Italy	18.1	0.5	3 382	-1.2	5.4	1.7	44.1	-1.8
Netherlands	10.5	3.4	304	2.6	34.7	0.8	12.4	-0.5
United Kingdom	9.9	-0.1	573	-0.5	17.3	0.4	26.4	-3.5

1. Annual growth rate from 2001 to 2011.
2. Swiss labour data are for 2008 and growth rate 2001-08.
Source: LEI calculation based on Eurostat for EU countries and BFS for Switzerland.

Table 5.10. **Share of dairy in manufacturing and labour productivity**
Based on turnover

	Share in manufacturing turnover			Labour productivity (€ 1 000 turnover per employee)		
	2001 (%)	2011 (%)	Growth[1] (%) 2001-11	2001	2008	Growth[1] (%) 2001-08
Switzerland[2]	2.2	1.1	-6.6	460	460	0.0
Austria	1.8	1.4	-2.3	423	378	-1.1
Germany	1.5	1.4	-0.8	538	562	0.5
Spain	1.7	2.3	2.6	280	296	0.6
France[a]	2.6	3.0	1.4	365	385	0.5
Italy	2.1	2.0	-0.6	315	313	-0.1
Netherlands	3.3	3.4	0.4	545	820	4.2
United Kingdom	1.4	1.7	2.2	259	262	0.1

1. Annual growth rate from 2001 to 2011.
2. Swiss labour data are for 2008 and growth rate 2001-08.
Source: LEI calculation based on Eurostat for EU countries and BFS for Switzerland.

products below world levels. The results are lower market shares in 2012 compared to 2000 for both imports and exports. Germany, France and the Netherlands are relatively large exporters as well as importers; all three are net-exporters. Switzerland is a net-exporter of dairy products in all years. Italy increased exports at a faster rate than world average and gained export share on the world market, but remained a large importer (Table 5.11).

Table 5.11. **Trade and market shares in processed dairy products**

	Export				Import			
	Export 2012 (Million USD)	Growth 2000-11 (%)	Market share 2000 (%)	Market share 2011 (%)	Import 2012 (Million USD)	Growth 2000-11 (%)	Market share 2000 (%)	Market share 2011 (%)
Switzerland	761	8.1	1.3	1.0	516	9.2	0.8	0.7
Austria	1 301	10.9	1.6	1.7	889	10.1	1.3	1.2
Germany	9 928	9.2	15.4	13.7	6 694	10.2	10.1	10.2
Spain	1 158	7.9	1.9	1.5	2 336	9.5	3.7	3.5
France	7 915	7.6	14.0	10.5	3 725	7.0	7.8	5.7
Italy	3 128	11.5	3.7	4.1	4 577	7.5	9.5	7.3
Netherlands	7 633	9.5	11.8	10.7	3 930	7.5	7.3	5.5
United Kingdom	1 735	6.6	3.6	2.4	3 838	7.7	6.8	5.4

Source: LEI calculation based on UNComtrade.

Manufacture of other food products

The overall competitiveness (O) of the Swiss "other food" processing industry (C108) is very strong compared to the benchmark countries (Figure 5.6). The main developments indicate that:

- The share of turnover of the "other food" products industry in total manufacture (S) increased in Switzerland, however growth was stronger in several benchmark countries. Thus, Switzerland is performing lower than average on this indicator.
- The growth of the real turnover (P) of the "other food" industry is rather strong and the highest of all selected countries.
- The growth of labour productivity (real turnover per employee (L) is strong and above average. In the period 2001 to 2008, the growth was 14% annually, far above the Netherlands with growth of 3% annually.
- The Relative Trade Advantage (T) index of Switzerland outperforms all other countries. Switzerland is a net-exporter of "other food" products.
- In addition, the performance of the export share on the world market (M) of Switzerland is the strongest.

Figure 5.6. **Competitiveness of the "other food products" industry**

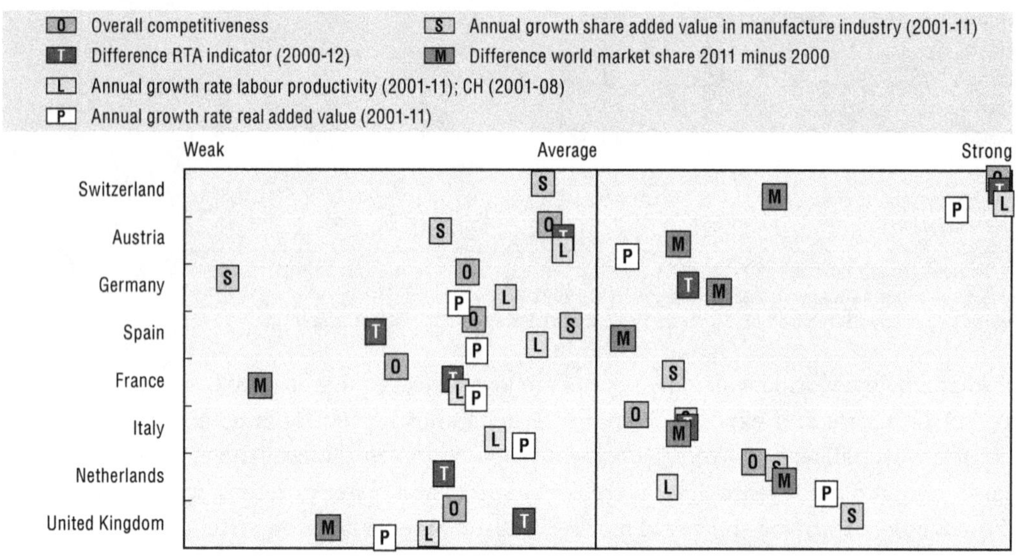

Source: LEI calculations based on Eurostat and FBS.

The turnover of the "other food" industry in Switzerland grew extremely fast: 10% annually, far above the levels of other countries that have also significant growth figures. The number of enterprises grew at a slower pace resulting in a strong growth of firm size as measured by average turnover. Swiss average turnover is 2.5 times higher than in the Netherlands and Germany (the second and third highest) and 15 times as high as in Italy (the lowest). This high rate of growth of turnover was accompanied by a decline in the number of employees in Switzerland (Table 5.12).

The sector "Manufacture of other food products", NACE C108 is a rather diverse sector and is subdivided in 7 sub-industries (Table 5.13). In Switzerland almost half of the turnover in 2011 was accounted for by cocoa and chocolate manufacturing. The share of

Table 5.12. **Structure of the "other food" industry in 2011**

	Turnover		Enterprises		Average turnover per enterprise		Employees	
	Billion (€)	Growth[1] (%)	Number	Growth[1] (%)	Million (€)	Growth[1] (%)	1 000	Growth[1] (%)
Switzerland[2]	24.4	10.0	471	3.1	51.9	6.7	15.1	-0.6
Austria	2.1	5.5	175	5.2	11.8	0.2	7.3	2.3
Germany	30.9	1.8	1 455	7.2	21.2	-5.0	101.5	1.3
Spain	10.8	3.6	2 480	-2.4	4.3	6.1	45.6	0.7
France	25.7	2.3	3 737	5.9	6.9	-3.4	79.5	3.2
Italy	19.9	4.0	5 443	1.7	3.7	2.3	57.6	3.0
Netherlands	11.3	6.8	521	5.5	21.7	1.2	22.4	4.1
United Kingdom	19.6	2.1	1 242	-0.2	15.8	2.3	92.3	2.2

1. Annual growth rate from 2001 to 2011.
2. Swiss labour data are for 2008 and growth rate 2001-08.
Source: LEI calculation based on Eurostat for EU countries and BFS for Switzerland.

Table 5.13. **Distribution of the sub industries of the "manufacture of other food products" industry**

NOGA/NACe	Description	2001		2011	
		Number (%)	Turnover (%)	Number (%)	Turnover (%)
108	Manufacture of other food products	100.0	100.0	100.0	100.0
1082	Manufacture of cocoa, chocolate and sugar confectionery	24.8	22.4	21.4	48.1
108201	Manufacture of cocoa, chocolate	11.8	19.6	13.6	46.8
108202	Manufacture of sugar confectionery	13.0	2.8	7.8	1.4
1083	Processing of tea and coffee	18.4	2.3	16.7	2.0
1084	Manufacture of condiments and seasonings	8.1	4.0	5.7	0.7
1085	Manufacture of prepared meals and dishes	17.0	0.9	7.2	2.5
1086	Manufacture of homogenised food preparations and dietetic food	8.1	2.1	9.3	0.7
1089; 1081	Manufacture of sugar and Manufacture of other food products n.e.c.	23.6	68.3	39.6	45.9

Source: LEI calculation based on BFS Mehrwertsteuer Schweiz.

this sub-industry grew strongly from 20% in 2001 to 47% in 2011 (Table 5.13). The turnover of the sub-industry cocoa and chocolate manufacturing grew extremely fast: 18% annually, and as such it is a driving element of the productivity and competitiveness increases in the whole sector.

The industry is largely based on imported raw materials and not those produced in the country itself, such as cocoa, tea or coffee. Hence, the raw materials base is also rather diverse, mainly based on imports or intermediate products from other industries. Sugar beet is grown in Switzerland, but the country is a net-importer of refined sugar. Self-sufficiency in refined sugar equivalents is between 50 to 60%.

The Swiss share of the "other food" industry in total manufacturing was already the highest of all selected countries in 2001 and increased further since. Swiss labour productivity (turnover per employee) in the sector is by far the highest of all countries and the growth of this indicator outperforms all other countries (Table 5.14).

Swiss exporters strengthened their position on the world market in recent years. While the world market grew by 13.9% in the period 2000-11, Swiss exports grew even faster by 16.4% (Table 5.15).

Table 5.14. **Share of "other food" industry in manufacturing and labour productivity**

Based on turnover

	Share in manufacturing turnover			Labour productivity (€ 1 000 turnover per employee)		
	2001 (%)	2011 (%)	Growth[1] (%)	2001	2011	Growth[1] (%)
Switzerland[2]	4.7	5.6	1.9	587	1 497	14.3
Austria	1.1	1.2	1.0	200	215	0.7
Germany	1.7	1.6	-1.0	275	250	-1.0
Spain	1.8	2.3	2.2	177	177	0.0
France	2.2	2.9	2.8	333	261	-2.4
Italy	1.6	2.2	2.9	300	264	-1.3
Netherlands	2.5	3.6	3.8	363	485	2.9
United Kingdom	2.1	3.3	4.5	207	147	-3.3

1. Annual growth rate from 2001 to 2011.
2. Swiss labour data are for 2008 and growth rate 2001-08.

Source: LEI calculation based on Eurostat for EU countries and BFS for Switzerland.

Table 5.15. **Trade and market shares in "other food"**

	Export				Import			
	Export 2012 (Million USD)	Growth 2000-11 (%)	Market share 2000 (%)	Market share 2011 (%)	Import 2012 (Million USD)	Growth 2000-11 (%)	Market share 2000 (%)	Market share 2011 (%)
Switzerland	3 896	16.4	1.7	2.5	1 760	11.3	1.3	1.2
Austria	1 672	13.2	1.0	1.1	2 138	12.5	1.3	1.3
Germany	14 093	13.0	8.4	8.8	10 308	11.1	7.4	6.5
Spain	3 072	10.9	2.2	1.9	4 030	12.9	2.7	2.8
France	7 519	7.9	7.4	4.7	8 078	10.8	6.0	5.1
Italy	5 323	12.8	3.1	3.1	4 270	12.3	2.6	2.6
Netherlands	10 896	13.8	6.2	7.0	7 030	14.6	3.8	4.7
United Kingdom	3 883	6.0	4.6	2.4	8 370	8.8	7.3	5.1

Source: LEI calculation based on UNComtrade.

The aforementioned developments are reflected in the trade indicators presented in Annex Figure A4.8. The Relative Export Advantage (RXA) index for Switzerland is above unity, which means that the country is a specialized exporter. The Relative Trade Advantage (RTA) indicator almost tripled from 0.3 in 1995 to 0.8 in 2012. France and the Netherlands also have a RXA above unity. These countries are also specialized importers of "other food", as the Relative Import Advantage (RMA) is also above unity. These countries are thus relatively specialized in trade of "other food" products. Austria, Spain, France and the United Kingdom are net-importers. Some have negative values for RTA. The data shows that the competitiveness of Switzerland "other food" products on the world market increased.

Manufacture of beverages

The overall competitiveness (O) of the Swiss beverages manufacturing industry (C11) is above average compared to the benchmark countries (Figure 5.7). The Austrian and Dutch industries are slightly stronger. The main developments indicate that:

- The share of the turnover of the beverages industry in total manufacture (S) is weak and declined in Switzerland.

- The growth in real turnover (P) in beverages manufacturing is above average; the Netherlands and Austria outperform Switzerland.
- The growth of labour productivity (real turnover per employee (L)) is below average.
- The Relative Trade Advantage (T) index of Switzerland outperforms all other countries. Nevertheless, Switzerland is a small net-importer of beverages.
- The performance of export share on the world market (M) of Switzerland is the strongest after Germany. Export from those countries grew twice as fast as the world average.

Figure 5.7. **Competitiveness of beverages manufacturing**

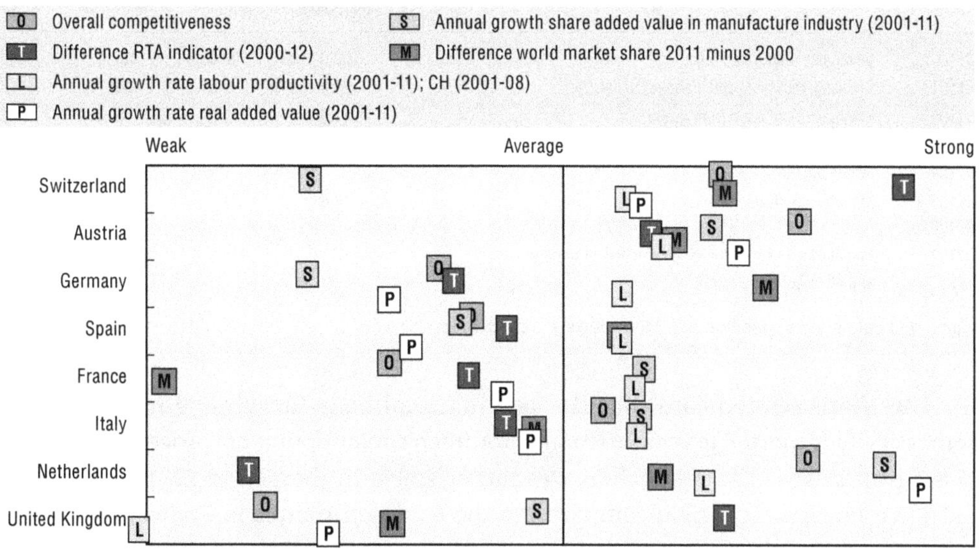

Source: LEI calculations based on Eurostat and FBS.

Swiss beverages manufacturing turnover showed modest growth (4.5%) between 2001 and 2011, higher than Germany and the United Kingdom (both -0.4%), lower than Austria and the Netherlands (10-12%). The average turnover per enterprise is in the range of the Southern European countries, but below the non-wine-producing EU benchmark countries (Netherlands, United Kingdom) (Table 5.16).

Table 5.16. **Structure of beverage manufacturing in 2011 (Switzerland 2008)**

	Turnover		Enterprises		Average turnover per enterprise		Employees	
	Billion (€)	Growth[1] (%)	Number	Growth[1] (%)	Million (€)	Growth[1] (%)	1'000	Growth[1] (%)
Switzerland[2]	3.2	4.5	367	1.6	8.6	2.8	7.1	2.1
Austria	4.9	9.7	365	3.2	13.4	6.3	9.0	-0.2
Germany	20.1	-0.4	2 019	0.1	10.0	-0.6	70.5	-1.2
Spain	15.8	1.6	4 557	0.2	3.5	1.4	47.8	-0.5
France	25.1	3.0	2 959	-1.7	8.5	4.7	44.1	-0.5
Italy	19.0	4.5	2 871	-0.4	6.6	4.9	35.9	-0.2
Netherlands	4.7	12.1	189	6.6	25.1	5.2	7.0	-3.0
UK	21.3	-0.4	1 033	3.3	20.6	-3.6		

1. Annual growth rate from 2001 to 2011.
2. Swiss labour data are for 2008 and growth rate 2001-08.

Source: LEI calculation based on Eurostat for EU countries and BFS for Switzerland.

Beverages manufacturing (NACE C110) is rather diverse and is subdivided in 5 sub-industries. Almost half of the turnover comes from manufacturing soft drinks and bottled waters. Second is beer production (34% of the turnover) and third wine production from grapes (15%). The latter sub-industry accounts for almost half of the enterprises and are –most probably – winegrowers, who produce wine from their own grapes (Table 5.17).

Table 5.17. **Distribution of the sub industries of beverage manufacturing (%)**

NACE	Description	2001		2011	
		Enterprises	Turnover	Enterprises	Turnover
C110	Beverages manufacturing	100.0	100.0	100.0	100.0
C1101	Distilling, rectifying and blending of spirits	29.4	9.5	20.1	8.0
C1102	Manufacture of wine from grape	41.9	12.3	45.9	15.1
C1103 & C1104	Manufacture of cider and other fruit wines & other non-distilled fermented beverages	3.8	1.2	4.4	1.2
C1105 & C1106	Manufacture of beer & malt	12.8	31.7	18.0	33.7
C1107	Manufacture of soft drinks; production of mineral waters and other bottled waters	12.1	45.3	11.5	42.1

Source: LEI calculation based on BFS Mehrwertsteuer Schweiz.

FAO statistics indicate that the raw material base for wine is mainly produced domestically. Imported grapes are mainly for fresh consumption or "processed" into ready-to-eat fruit salads. The production was rather stable in the period 1991 to 2011, and, at 130 000 tonnes, is negligible compared to the 6 million tonnes in France, Italy or Spain. Austria produces twice as much as Switzerland. The price of Swiss grapes is high compared to the EU benchmark countries (Annex Figure 5.A1.5). For beer one of the inputs is barley malt, which is imported duty free.

The share of turnover in beverages manufacturing in total manufacturing is declining (-3.2%) on the same level as in Germany. In all other benchmark countries the share increases: relatively high in the Netherlands and Austria (Table 5.18).

Table 5.18. **Share of beverage manufacturing in total manufacturing and labour productivity**

	Share in manufacturing turnover			Labour productivity (€ 1 000 turnover per employee)		
	2001 (%)	2011 (%)	Growth[1] (%)	2001	2011	Growth[1] (%)
Switzerland[2]	1.0	0.7	-3.2	324	322	-0.1
Austria	1.7	2.8	5.0	205	415	7.3
Germany	1.4	1.0	-3.2	253	234	-0.8
Spain	3.3	3.4	0.2	269	247	-0.8
France	2.0	2.8	3.5	384	460	1.8
Italy	1.5	2.1	3.4	324	405	2.3
Netherlands	0.7	1.5	8.9	148	649	15.9
United Kingdom	3.0	3.6	1.9	382		

1. Annual growth rate from 2001 to 2011.
2. Swiss labour data are for 2008 and growth rate 2001-08.

Source: LEI calculation based on Eurostat for EU countries and BFS for Switzerland.

Swiss beverages manufacturing is quite competitive: their export share on the world market grew annually by 28%, far above the world average of 11%. Export of soft drinks and bottled water accounts for the largest share of exports. At the same time, imports grew at a slower pace resulting in a lower import marker share. Switzerland remained a (very small) net-importer of beverages just like Germany. All other selected countries are net-exporters. The two leading exporters France and the United Kingdom lost share on the export market: France from a market share of 23% in 2000 to 17% in 2011 and the United Kingdom from 14 to 11% (Table 5.19).

Table 5.19. **Trade and market shares in beverages products**

	Export				Import			
	Export 2012 (Million USD)	Growth 2000-11 (%)	Market share 2000 (%)	Market share 2011 (%)	Import 2012 (Million USD)	Growth 2000-11 (%)	Market share 2000 (%)	Market share 2011 (%)
Switzerland	1 851	27.9	0.3	1.7	1 901	7.6	2.5	2.0
Austria	2 394	13.4	1.8	2.4	734	10.9	0.7	0.8
Germany	6 098	14.0	4.3	6.2	7 822	9.5	8.7	8.5
Spain	4 665	10.0	4.4	4.3	2 280	8.0	3.1	2.6
France	17 856	7.4	23.1	17.3	3 915	9.7	4.3	4.3
Italy	8 053	9.3	8.8	8.0	1 940	8.2	2.5	2.1
Netherlands	5 194	11.1	4.8	5.2	3 993	12.8	3.1	4.2
United Kingdom	10 897	8.0	13.6	10.8	8 397	6.4	11.9	8.5

Source: LEI calculation based on UNComtrade.

Switzerland has become a specialized exporter of beverages between 1995 and 2012. The Relative Export advantage index is above unity in 2012 and the country remained a specialized beverages importer. The Relative Import Advantage Index is above unity. Both developments together resulted in an improvement of the Relative Trade Advantage index from -1.5 in 1995 to -0.4 in 2012 (Annex Figure 5.A1.9).

Conclusions

The assessment of revealed competitiveness of the Swiss food and beverage sector shows a mixed picture. While there are some very strong performing sub-sectors, most of the food industry performs weakly compared to competitors in EU benchmark countries. The strong performing "other food" manufacturing sub-sector accounts for around 60% of industry turnover and half of exports. Almost half of the turnover of this sub-sector comes from the manufacturing of cocoa and chocolate (2011). "Other food" turnover grew annually by 10%: almost twice as fast as the overall food & beverages industry (5.8%) in the period 2001-2011. Another strong sub-sector is beverages, accounting for 12% of industry exports. The oils and fats industry also performs well, and is strongly linked to the "other food" industry as supplier of intermediate products to produce condiments, seasonings or meals. In these strongly performing sub-sectors, a major part of the raw materials is imported or non-agricultural (mineral water).

On the other hand, the weakest sectors, meat, dairy and animal feed are those largely based on domestic raw materials. These industries have to pay relatively high prices for their agricultural inputs as most of these prices are well above EU price levels. Additionally, these less competitive sectors have a relatively low increase of labour productivity and are relatively labour intensive. This contrasts with more competitive Swiss sectors (see above) which recorded much higher growth in labour productivity.

The current import protection regime and agricultural policies are inhibiting a more dynamic participation in global and regional value chains for agro-food. OECD (2013) shows that successful participation in such value chains, where specialized businesses at each stage of production add value to the product before it enters the final consumer market, requires unencumbered access to the best inputs at the lowest prices as well as regulations and technical standards that facilitate exchange of semi-processed and finished products with partner countries. Competiveness of the agro-food industry can be enhanced by more transparent and less regulated markets both upstream and downstream (BAK Basel 2014).

Structural change in the Swiss food industry will continue, and will require exploitation of scale economies and identifying niche markets. Forward looking reforms of agricultural policies will be necessary to support the development of a more market oriented commercial farming sector such that it can contribute to increasing the competitiveness of those Swiss food industries that are mainly based on domestic agricultural raw materials (EAER 2014). Farmers and food consumers would also benefit from improved transparency and more competition in the downstream sector, including the retail level (Hediger, W., El Benni, N., 2014).

Notes

1. While this study aims to be comprehensive in its coverage of agri-food sectors, it does not consider the retail sector in the analysis.
2. The applied methods and indicators are discussed more thoroughly in the Annex to this chapter.
3. A potential flaw of the RXA is that re-exports might suggest high competitiveness of one industry. These transit activities might be influenced by a good performance of another sector i.e. logistics or by beneficial natural and infrastructural conditions like sea or airports. These features are less relevant in the Swiss context however.
4. UNComtrade database provides all trade data; Eurostat-SBS database for all economic data for the EU countries; FBS "Mehrwertsteuer" and labour statistics as economic data sources for Switzerland; FAOstat for production, price and self-sufficiency of raw materials. A major handicap for assessing the Swiss food competitiveness is non-availability of value added data. Instead turnover has been taken as a proxy. Only for the recent years 2009-2011 data are comparable between Switzerland and EU-countries.
5. The series of graphs in this chapter presents the values of the indicators as z-scores, which scales all variables to a distribution with a mean of 0 and a standard deviation of 1. A "Z-score" of 0 means an average score, a negative Z-score a weak score and a positive Z-score means strong.

ANNEX 5.A1

Price comparisons and trade indicators

Figure 5.A1.1. **Farmers' prices for pig meat, 1991-2011**

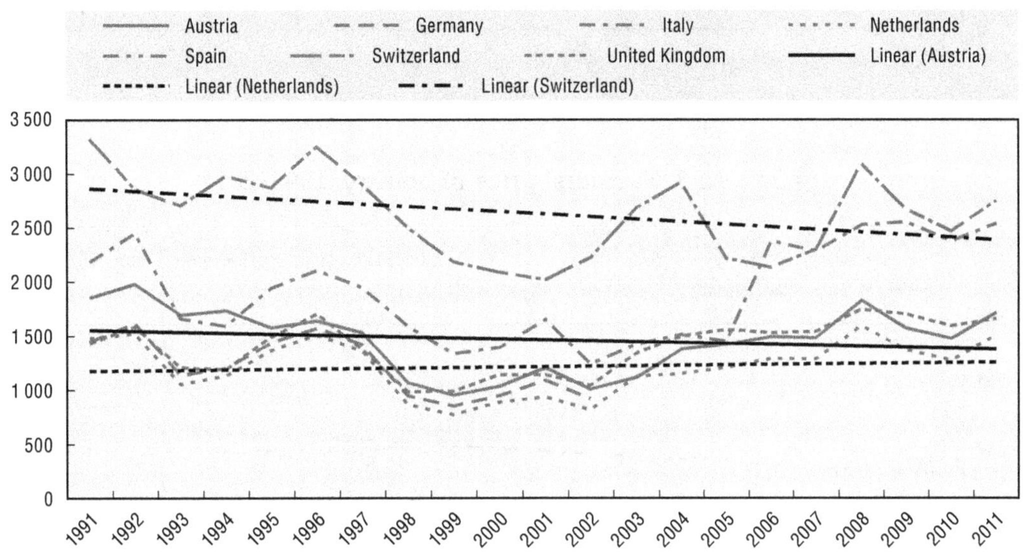

Source: LEI calculations based on FAOStat.

Figure 5.A1.2. **Farmers' price of cattle meat, 1991-2011**

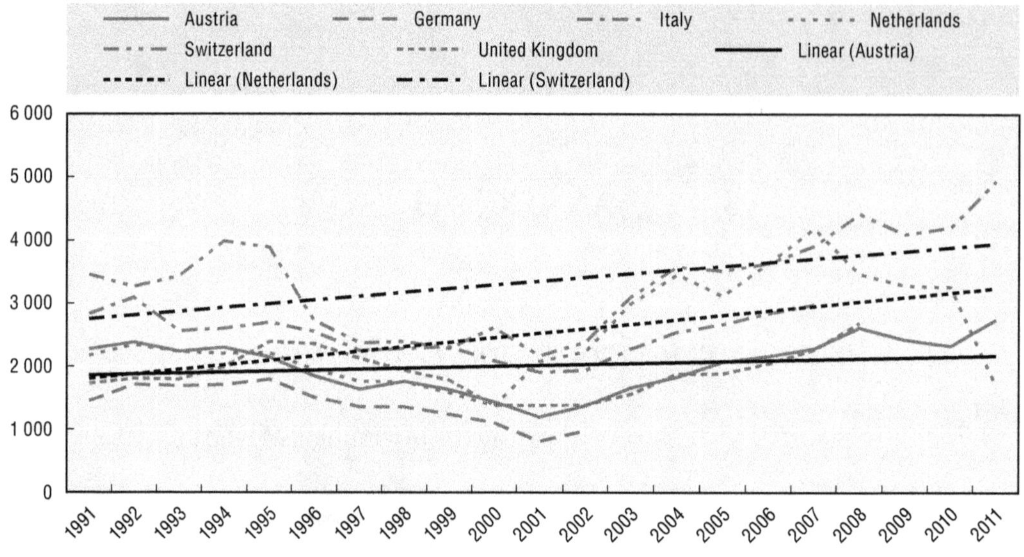

Source: LEI calculations based on FAOStat.

Figure 5.A1.3. **Farmers' price of poultry, 1991-2011**

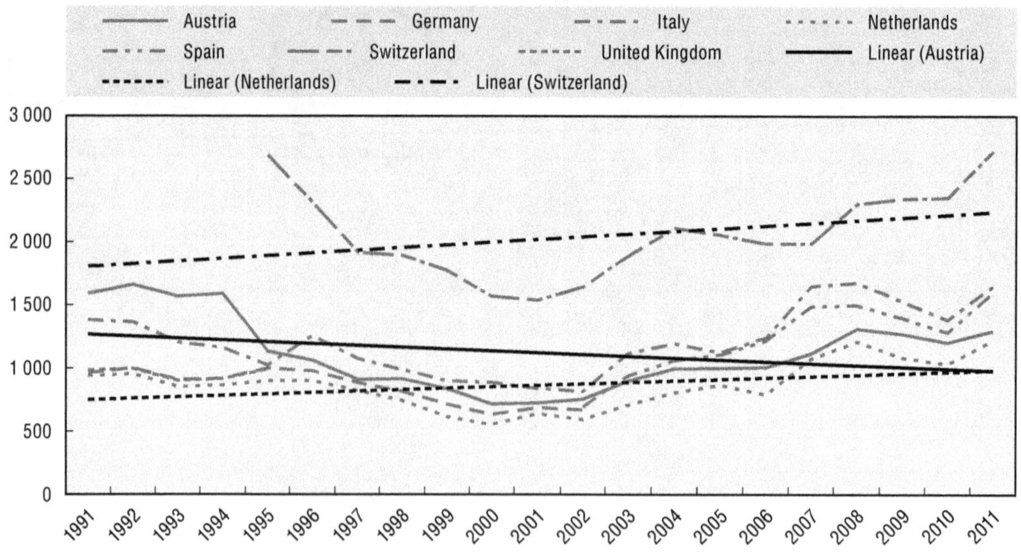

Source: LEI calculations based on FAOStat.

Figure 5.A1.4. **Farmers' prices of milk, 1991-2011**

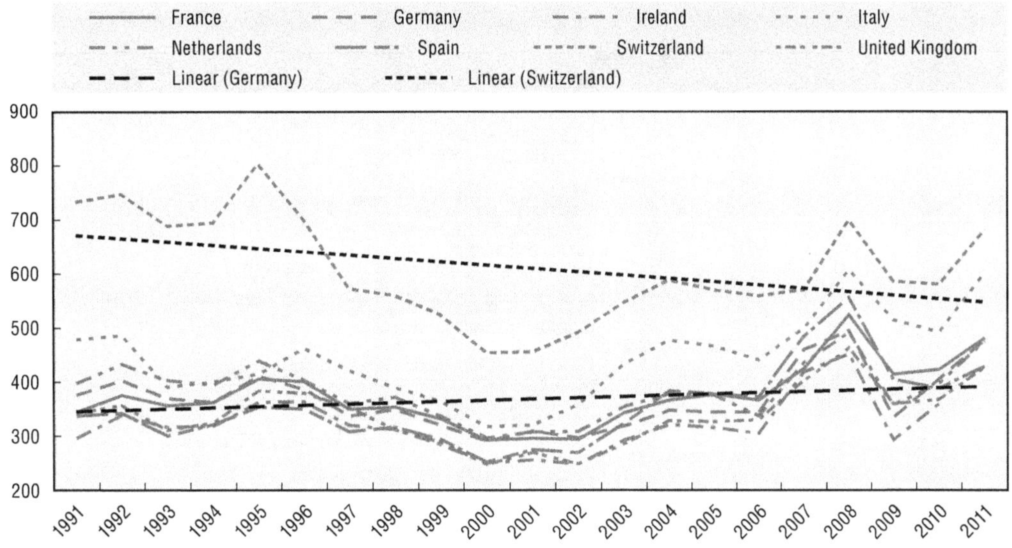

Source: LEI calculation based on FAOStat.

Figure 5.A1.5. **Prices of grapes in USD/ton**

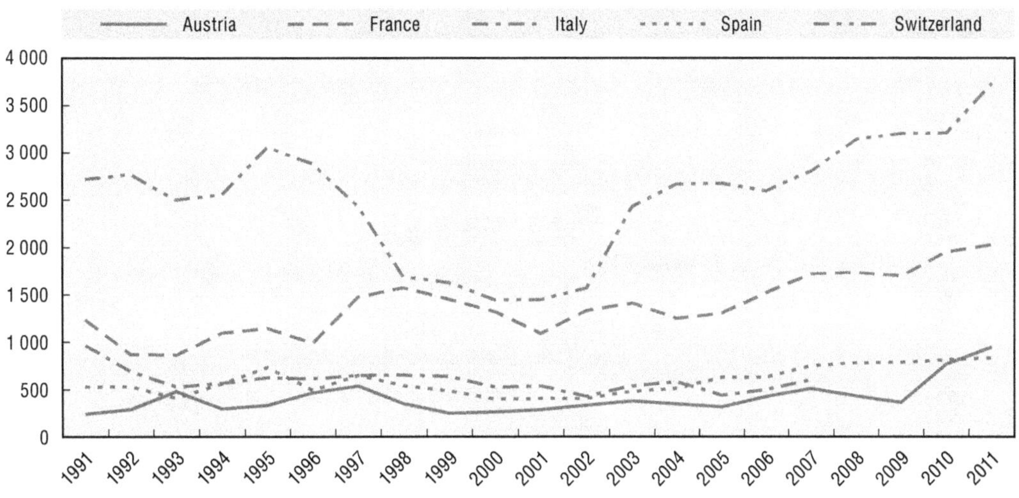

Source: LEI calculations based on FAOStat.

Figure 5.A1.6. **Trade indicators of oils meat processing**

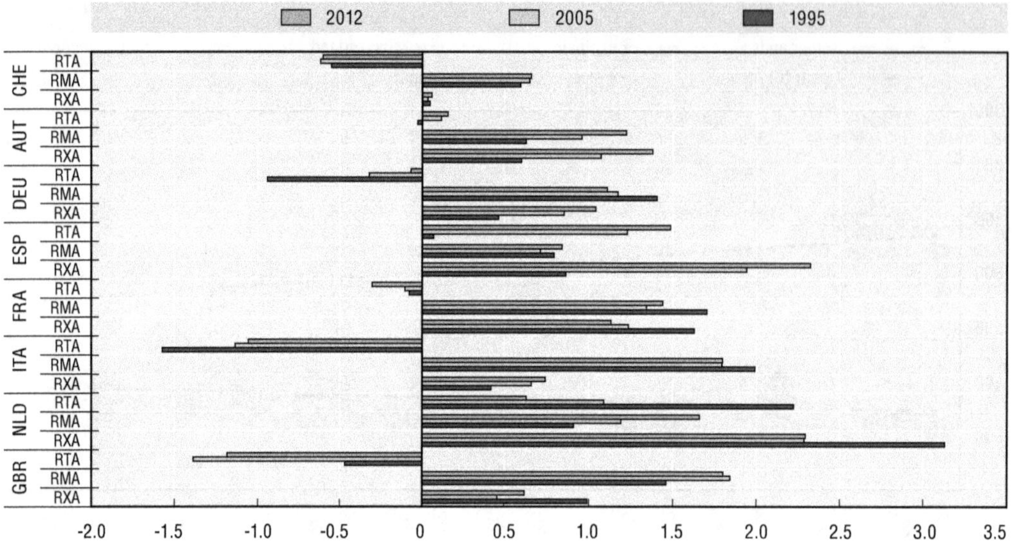

Note: RTA, RMA, RXA denote relative trade/import/export advantage indexes and are defined in Annex 5.1.
Source: LEI calculations based on UNComtrade.

Figure 5.A1.7. **Dairy trade indicators**

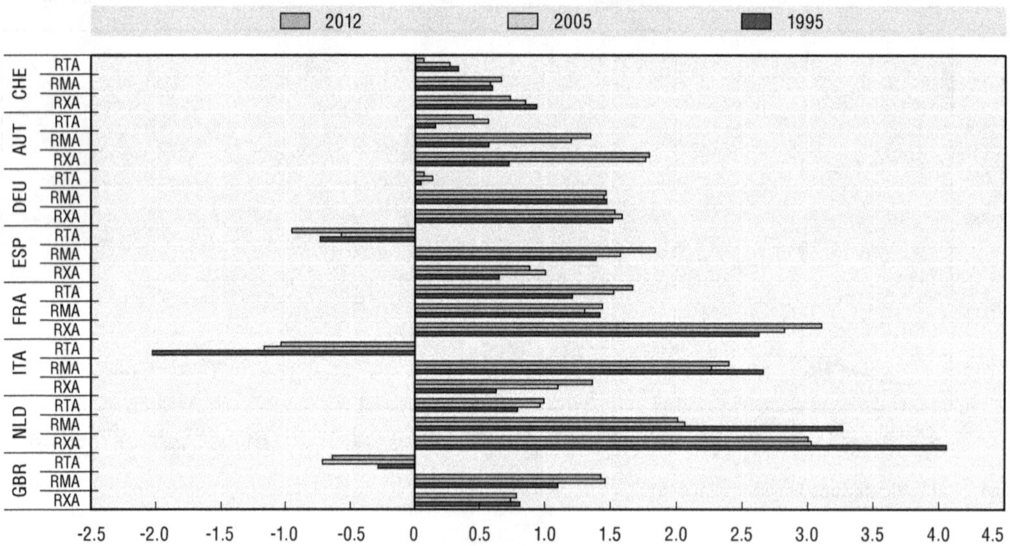

Note: RTA, RMA, RXA denote relative trade/import/export advantage indexes and are defined in Annex 5.1.
Source: LEI calculations based on UNComtrade.

Figure 5.A1.8. **Trade indicators of the "other food" industry**

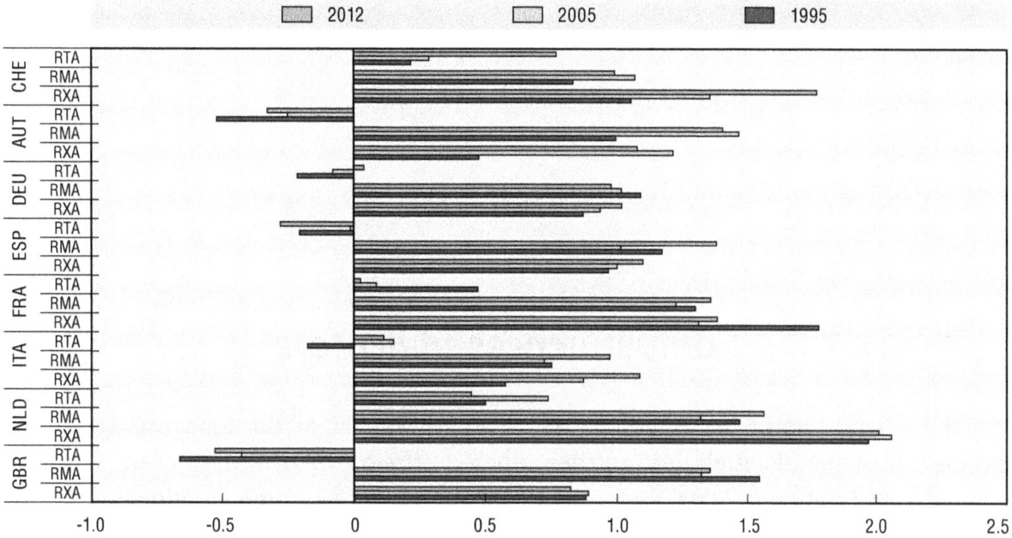

Note: RTA, RMA, RXA denote relative trade/import/export advantage indexes and are defined in Annex 5.1.
Source: LEI calculations based on UNComtrade.

Figure 5.A1.9. **Trade indicators of the beverage industry**

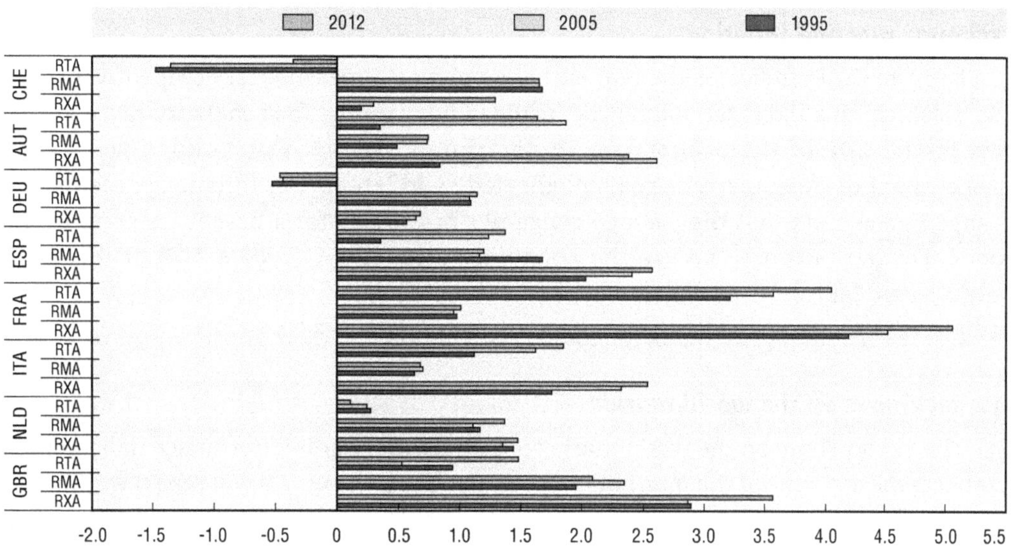

Note: RTA, RMA, RXA denote relative trade/import/export advantage indexes and are defined in Annex 5.1.
Source: LEI calculations based on UNComtrade.

ANNEX 5.B1

Competiveness indicators

The starting point for assessing the competitiveness of the food industry is the approach of Wijnands et al. used in their study on the competitiveness of the European food industry (Wijnands, 2008; Wijnands, 2007). In this annex, some additional indicators are discussed that might be used in assessing the competitiveness. The overview below is far from exhaustive. Furthermore, a distinction is made between trade measures and business economic performance measures of competitiveness.

Trade related indicators

Exchange rate and inflation

Latruffe (2010) indicates the real exchange rate as a measure for competitiveness. In this research, this indicator will not be included because the food industry takes a small part in the GDP of the economies. To determine the real value added we use the development of consumer prices also indicated as inflation. The inflation measures the change in the costs that the average consumer has to pay for a basket for services and goods. For our purpose, we use the consumer price index (2005 = 100) of the World Development Indicators database.

CP_{ct} is consumer price index for country c in period t

Market shares on the world market

The export share on the world market is a straightforward performance indicator and it reflects the outcome of the international competitive process. We will take the difference between two periods of a country's export share on the world market. The growth we measured is the change and not an annual growth rate between two periods, as we will propose for other indicators. Growth rates between two periods have a strong flaw. Very small exporters can have large growth rates, but remaining small exporters. Even with small growth rates, large exporters will have a larger impact on the market. The definition of this indicator reflects the strong interdependency between the exports of the different countries. By taking the absolute deviation, the real impact on the world market is taken into account. Furthermore, the total sum of all changes is by definition zero. Table 5.B1.1 gives an example of the discussion above taken from (Wijnands, 2007).

(1) $GES_{ict} = MS_{ict_1} - MS_{ict_2}$

GES_{ict} Growth export share on the world market for industry i for country c between period t_1 and t_2

MS_{ict} Export share on the world market for industry i for country c in period t

c Selected country

i Selected industry according to classification of NACE

t Selected year

(2) $MS_{ict} = X_{ict} / X_{iwt}$

X_{ict} The export value of industry i, country c in period t.

X_{iwt} The export value of industry i of the world (as a whole) in period t.

Table 5.B1.1. **Example of impact of indicators and market shares development.**

	Market share (%)			
	1996-98	2002-04	Deviation	Growth
Country A	1	2	1	100%
Country B	50	51	1	2%
Country C	20	20	0	0%
Country D	29	27	-2	-7%

Revealed comparative advantage indices

The relative importance of an industry in the total trade is usually measured by the Revealed Comparative Advantage (RCA) or Balassa index or specialisation index (Fertö and Hubbard, 2003; Latruffe, 2010; Wijnands, 2008). If it is related to the export, it measures the export share of a product of one country in the total export of the world relative to the country's export share in the world of all products. The relative export advantage index is as follows:

(3) $RXA_{ict} = \dfrac{X_{ict} / X_{iwt}}{XT_{ct} / XT_{wt}}$ Export value of specific industry i from country c in period t.

RXA_{ict} the relative export advantage index for industry i, country c in period t.

X_{ict} The export value of industry i, country c in period t.

XT_{ct} The total export value of all industries of country c in period t.

XT_{wt} The total export value of all industries in the world in period t.

The total export value of all industries from one country is the total of all export: unprocessed or processed agriculture commodities, or industrial products or services.

The flaw of this index is that re-export might suggest high competitiveness of one industry. These transit activities might be influenced by a good performance of another sector *i.e.* logistics or by beneficial natural and infrastructural conditions like sea or airports.

A RXA index of 1 indicates that a country is equally specialised as the total world exports. A level below 1 means relatively unspecialised and above 1 relatively specialised. The latter indicates an export advantage, as relative more is exported than the world average. In fact it indicates the export focus of an industry and is therefore externally oriented. Again the annual growth between the first and last time period will be used. The index is only relevant for exporting industries.

The opposite of the relative export advantage index is the relative import advantage index:

(4) $RMA_{ict} = \frac{M_{ict}/M_{iwt}}{MT_{ct}/MT_{wt}}$ import value of specific industry i from country c in period t.

RMA_{ict} The relative import advantage index for industry i, country c in period t.

M_{ict} The import value of industry i of country c or of the world w in total in period t.

MT_{ct} The import value of all industry i of country c or of the world w in total in period t.

The interpretation of the index is reversed from that of RXA. A value below unity (= 1) shows that country imports relatively less than the world average and can be indicated as a competitive advantage; a value above unity indicates a relative higher import level.

High levels or re-export of products, due to comparative advantage of other sectors or country's location, might explain a high value.

The Relative Trade Advantage index is defined by Scott and Vollrath as difference between the RXA and RMA (Scott and Vollrath, 1992).

(5) $RTA_{ict} = RXA_{ict} - RMA_{ict}$

A positive RTA indicates a competitive advantage: the exports exceed the imports. Negative values signify competitive disadvantages (Scott and Vollrath, 1992).

The advantage of these indices is the simplicity to calculate these indicators based on an available and well accessible database. In this report, the values of all three indices will be presented. As metrics in the assessment of the competiveness, the absolute growth between 2 periods of the Relative Trade Advantage will be used as this index summarizes the export and import developments. The is index has advantages above the indices based on either export or imports (Frohberg and Hartmanm, 1997). This indicator is modification of the approach of Wijnands et al., 2008. A positive growth indicates an increase in domestic supply of that product, meaning that the industry gains competitiveness compared to other countries.

Other indices based on trade

Several other indicators related to international trade are available such as the Net Trade Ratio that expresses the ratio between imports and exports of a country or the Grubel-Loyd intra-industry trade index, Porter-adapted index of RXA or the Dunning adapted RXA. Furthermore, several modifications of the indices mentioned above are discussed in the literature (Frohberg and Hartmanm, 1997; Gellynck, 2002; Latruffe, 2010). We do not consider these indices because above we already mentioned the export and import advantage indices whose interpretation is less complicated in terms of competitiveness. The Porter and Dunning indices include outward an inbound production. We do not consider these indices as we will present below, because as we are using data from national accounts that includes only domestic production.

Economic indicators

The selected indicators for quantifying the industry's competitiveness are taken from Wijnands et al., 2008. Due to insufficient data turnover is used, as value added was not available, unless otherwise stated. For the period 2009-11 a comparison based on value added will be provided. Therefore, in this section both indicators are indicated by the same acronym.

Turnover versus Real value added

Creating added value is an important economic indicator. It is related to the industrial dynamism. Total value added is not only based on the production factor labour but also on the production factor capital and land. In study turnover is used as proxy. The indicator of growth is taken, so that countries can be compared easily. Annual growth in real value added of the food industry (or subsector) is used. Their growth is taken as an indicator, so that countries can be compared despite differences in PPP.

To derive the real value added at factor costs/turnover, the nominal value added/ turnover is deflated by the consumer price index.

$$(6) \quad RVA_{ict} = \frac{VA_{ict}}{CP_{ct}}$$

RVA_{ict} Real value added/ turnover for industry i in in country c for period t

VA_{ict} Nominal value added/turnover for industry i in country c for period t

CP_{ct} Consumer price indicator for country c in period t

Real value added versus turnover shares

The importance of a specific sub-industry is derived from its share in the food industry. A growth in the share reflects a competitive advantage. The industry is then able to attract resources for their production. This reflects the competition for production factors (labour and/or capital) between different industries within a country.

The food industry is used for comparison, if a sub-sector of the food industry, e.g. dairy processing, is evaluated. Where the food industry as whole is evaluated, the manufacturing industry has been used. The metrics is the growth of the share of the specific industry in the food industry. A positive growth shows a better than average performance than the food industry as a whole.

$$(7) \quad SRVA_{ict} = \frac{RVA_{ict}}{RVA_{mct}}$$

$SRVA_{it}$ Share of the real value added/turnover for industry i in total manufacture industry (m) in country c for period t

m Manufacture industry as a whole

Labour productivity based on real value added viz. turnover

Labour productivity affects prices in the market. Growth of labour productivity improves industrial competitiveness in international markets. Labour productivity is often seen as a crucial determinant of competitiveness. The labour productivity is the real value added viz. turnover divided by the number of employees. This indicator cannot be compared between different countries due to different levels of Purchasing Power Parities. As we take the growth of the labour productivity, the indices of different countries can be compared. This indicator can be seen as measurement of the potential competitiveness.

$$(8) \quad RLP_{ict} = \frac{RVA_{ict}}{E_{ict}}$$

RLP_{ict} is real labour productivity for industry i in country c for period t

E_{ict} is number of employees in industry i in country c for period t

Exchange rates

All indicators are growth percentages. Growth percentages are not influenced by exchange rates, so they can be calculated in the original currency. The nominal values in the descriptive parts are all converted to Euros with the exchange rate as mentioned by Eurostat.

Competitiveness assessment

Annual growth rates of the indices

According to Porter sustainable competitive advantage is the fundamental source for above-average performance in the long run (Porter, 1980; Porter, 1990). In line with Porter's viewpoints, competitiveness of the food industry is defined as the sustained ability to achieve profitable gain and market share in domestic and export markets in which the industry is active. Annual growth rates (except for market shares on the world market and for the relative trade advantage index) between 2 periods are used as indicators. High growth rates indicate high ex-post performance, compared to other industries of a particular country.

Comparison of indicators and overall competition

The food industries will be benchmarked against a number of selected countries. The benchmark will be presented for each sub-industry and the food industry as a whole.

The above-mentioned indicators have different scales. To compare the different scales the values will be standardised. Calculations are:

X_i is observation i=1,n (ic number of countries)

$$\overline{X} = \sum X_i / n$$

$$s = \sqrt{\frac{\sum (X_i - \overline{X})^2}{n}}$$

$$z_i = \frac{X_i - \overline{X}}{s}$$

All variables will become the same dimension (average and standard deviation) and can than easily presented in one figure. Furthermore the mean of these values can be calculated as an indication of the overall competitiveness. In this case, the implicit assumption was that the weight or importance of each indicator is equal. It is possible to impose different weights for each indicator. However, no empirical evidence is currently available for different weights.

However, this method also has a disadvantage. The standard scores depend on the number of the countries and the levels of indicators in the sample: the standard scores are not fixed. It is a fact a benchmark, and if the benchmark countries or the level of one indicator in one country changes, the position of the countries will also change. It is a relative position.

References

BAK Basel (2014), *Landwirtschaft – Beschaffungsseite, Vorleistungsstrukturen und Kosten der Vorleistungen*, BAK Basel.

BFS (2013), Die *Mehrwertsteuer in der Schweiz 2010-11, Resultate und Kommentare, Bundesamt für Statistik* (BFS), Neuchâtel.

EAER (Federal Department of the Economic affairs, Education and Research) (2014), Une politique industrielle pour la Suisse, Rapport du 16 avril 2014 faisant suite au postulat Bischof (11.3461), *www.news.admin.ch/NSBSubscriber/message/attachments/34529.pdf*.

EC (2008), NACE Rev. 2 – Statistical classification of economic activites in the European Community, European Commission: Office for Official Publications of the European Communities, Luxembourg, pp. 363.

Fertö, I., L.J. Hubbard (2003), "Revealed Comparative Advantage and Competitiveness in Hungarian Agri–Food Sectors", *The World Economy* 26:247-259.

Frohberg K., M. Hartmanm (1997), *Comparing measures of competitveness, Institute of Agricultural Development in Central and Eastern Europe*, Halle.

FSO (2008), NOGA 2008 General Classification of Economic Activities, Federal Statistical Office, Neuchâtel.

Gellynck, X. (2002), "Changing Environment and Competitiveness in the Food Industry", Faculteit Economie en Bedrijfskunde, University Gent, Gent.

Hediger W., N. El Benni (2014), Schlussbericht zum Forschungsprojekt "Wettbewerbsfähigkeit Landwirtschaft – Nachgelagerte Industrien", Hochschule für Technik und Wirtschaft (HTW) Chur.

Jarrett, P. and C. Moeser (2013), "The Agri-food Situation and Policies in Switzerland", *OECD Economics Department Working Paper*, No. 1086, September, OECD Publishing, Paris, *http://dx.doi.org/10.1787/5k40d6ccd1jg-en*.

Krugman, P.R. and M. Obstfeld (2006), *International Economics: Theory and Policy*, Pearson/Addison-Wesley, Boston.

Krugman, P.R. and M. Obstfield (1988), *International economics:theory and implications*. 6th ed. Addison-Wesley, Boston.

Latruffe, L. (2010), "Competitiveness, Productivity and Efficiency in the Agricultural and Agri-food Sectors", *OECD Food, Agriculture and Fisheries Papers*, No. 30, OECD Publishing, Paris, *http://dx.doi.org/10.1787/5km91nkdt6d6-en*.

OECD (2013), *Interconnected Economies: Benefiting from Global Value Chains*, OECD Publishing, Paris, *http://dx.doi.org/10.1787/9789264189560-en*.

O'Mahoney, M. and B. van Ark (2003), "EU-productivity and competitiveness: an industry perspective", Office for Official Publication of the European Community, Luxembourg, pp. 273.

Porter, M.E. (1990), *The Competitive Advantage of Nations*, The MacMillam Press Ltd, London.

Porter, M.E. (1980), *Competitive strategy: Techniques for Analyzing Industries and Competitiors*, The Free Press, New York.

Sagheer S., S.S. Yadav and S.G. Deshmukh (2009), "Developing a conceptual framework for assessing competitiveness of India's agrifood chain", *International Journal of Emerging Markets* 4:137-159.

Scott L. and T. Vollrath (1992), *Global Competitive Advantages and Overall Bilateral Complementarity in Agriculture: A Statistical Review*, Economic Research Service, US. Department of Agriculture, Washington.

Siggel, E. (2006), "International competitiveness and comparative advantage: a survey and a proposal for measurement", *Journal of Industry, Competition and Trade* 6:137-159.

Spence, A.M. and H.A. Hazard (1998), *International Competitiveness*, Ballinger Publishing Company, Cambridge, MA.

Thompson, A.A. and A.J. Strickland (2003), *Strategic management. Concepts and Cases*, McGraw Hill, New York.

Wijnands, J.H.M., H.J. Bremmers, B.M.J. van der Meulen and K.J. Poppe (2008), "An economic and legal assessment of the EU food industry's competitiveness", *Agribusiness* 24:417-439, DOI:10.1002/agr.20167.

Wijnands, J.H.M., B.M.J. van der Meulen and K.J. Poppe (2007), *Competitiveness of the European Food Industry : An economic and legal assessment*, Office for Official Publications of the European Communities, Luxembourg, pp. 328.

Wright, P., M. Kroll and J. Parnell (1998), *Strategic Management: Concepts and Cases*, Prentice Hall (4th ed.).

ORGANISATION FOR ECONOMIC CO-OPERATION AND DEVELOPMENT

The OECD is a unique forum where governments work together to address the economic, social and environmental challenges of globalisation. The OECD is also at the forefront of efforts to understand and to help governments respond to new developments and concerns, such as corporate governance, the information economy and the challenges of an ageing population. The Organisation provides a setting where governments can compare policy experiences, seek answers to common problems, identify good practice and work to co-ordinate domestic and international policies.

The OECD member countries are: Australia, Austria, Belgium, Canada, Chile, the Czech Republic, Denmark, Estonia, Finland, France, Germany, Greece, Hungary, Iceland, Ireland, Israel, Italy, Japan, Korea, Luxembourg, Mexico, the Netherlands, New Zealand, Norway, Poland, Portugal, the Slovak Republic, Slovenia, Spain, Sweden, Switzerland, Turkey, the United Kingdom and the United States. The European Commission takes part in the work of the OECD.

OECD Publishing disseminates widely the results of the Organisation's statistics gathering and research on economic, social and environmental issues, as well as the conventions, guidelines and standards agreed by its members.

OECD PUBLISHING, 2, rue André-Pascal, 75775 PARIS CEDEX 16
(51 2014 09 1 P) ISBN 978-92-64-22654-8 – 2015-17

www.ingramcontent.com/pod-product-compliance
Lightning Source LLC
Chambersburg PA
CBHW061218240926
51846CB00006B/1266